The Houghton Mifflin Series in Statistics
under the Editorship of Herman Chernoff

Leo Breiman
Probability and Stochastic Processes / With a View Toward Applications
Statistics / With a View Toward Applications

Paul G. Hoel, Sidney C. Port, and Charles J. Stone
Introduction to Probability Theory
Introduction to Statistical Theory
Introduction to Stochastic Processes

Paul F. Lazarsfeld and Neil W. Henry
Latent Structure Analysis

Gottfried E. Noether
Introduction to Statistics—A Fresh Approach

Y. S. Chow, Herbert Robbins, and David Siegmund
Great Expectations: The Theory of Optimal Stopping

I. Richard Savage
Statistics: Uncertainty and Behavior

Introduction

Gottfried E. Noether

University of Connecticut

to Statistics

A Fresh Approach

Houghton Mifflin Company
Boston

New York
Atlanta
Geneva, Illinois
Dallas
Palo Alto

Library of Congress Catalog Card Number: 77-135750

ISBN: 0-395-05001-4

To L.

Editor's Introduction

The introductory course in statistics has always represented a serious pedagogic problem. While it will be terminal for many students, for others it will provide a basis for studying specialized methods within their major fields. The main function of such a course is to introduce them to variability, uncertainty, and some common statistical methods of drawing inferences from observed data. Such a course should be intellectually stimulating; the ability to reason statistically should be more crucial to the success of a student than his mathematical ability.

Traditional textbooks do not generally serve the needs adequately. In most cases the basic statistical ideas do not appear until very late. Valuable time is first devoted to the intellectually dull task of organizing and graphing data, and to interminable computations of means, modes, medians, and variances. Sometimes effort is still devoted to rather pointless discussions of skewness, kurtosis, and geometric means. Students with little interest in mathematics find a week or more of combinatorics frustrating if not mystifying. Finally one-quarter or one-semester courses which attempt to cover most of the important methods in statistics run out of time or become dull compendia of formulae. It is preferable to concentrate on basic ideas and a few methods and to let the applied departments introduce specialized techniques when needed at a small cost of time.

What are called for are innovative approaches which avoid these pitfalls and are efficient in exhibiting many basic ideas of inference in contexts which are intellectually stimulating. Professor Noether has presented us with such a book. The main technique is the exploitation of nonparametric methods. This book is not a study of nonparametric methods in statistics. Rather, it is the effective use of these methods to illustrate statistical ideas.

What are the advantages of this approach? The basic ideas of inference appear at the beginning. The methods developed are easy to apply and require a minimal amount of computation. The methods are simple in principle; the common sense logic behind them is easy to perceive and to explain. The probability background used in most of the illustrations is minimal, requiring often no more than easy applications of the binomial distribution and its tables. Although no attempt is made at complete coverage, this vehicle is remarkably effective in covering many basic ideas with a few methods and illustrations. Finally, there are the advantages of the robustness and of the wide applicability of the nonparametric methods discussed.

This is a well conceived, carefully thought out, and well written book. I fully anticipate that Professor Noether's success with a preliminary version will be shared by many teachers who use this book.

Herman Chernoff

Preface

In writing this book I have tried to avoid two dangers that seem ever present when teaching a first course in statistics to non-majors. At one extreme students are overwhelmed by probability theory. At the other extreme they are bored by what are to them endless and mostly meaningless computations. In the present book, topics in the theory of probability and in descriptive statistics are held to bare essentials.

The aim of this book is to allow students to concentrate on basic ideas without becoming involved in technical and computational detail. Accordingly, such concepts as estimation and hypothesis testing are discussed in terms of the binomial model, which is much simpler than the normal model. One-, two-, and k-sample problems are solved nonparametrically before the student is introduced to t-tests and the analysis of variance; rank correlation precedes least square regression and product moment correlation. This arrangement is based on my conviction and experience that nonparametric methods are not only safer—particularly in the hands of beginners—but are also conceptually and computationally simpler than the corresponding normal theory methods.

A reasonably good high school background in mathematics is sufficient for the book.

The book is intended chiefly for a one-semester or two-quarter course; but it provides sufficient subject matter for a year course. A flexible organization permits the instructor of a one-semester course to choose either a purely nonparametric or a combination nonparametric-normal theory approach. I have covered the contents of Chapters 1 through 16 in a one-semester course, meeting three times a week, by omitting all or most of Chapter 4 on random variables.

A possible compromise between complete omission and complete inclusion of Chapter 4 in a one-semester course is to take up, rather briefly, the ideas of Sections 21 through 24 before going on to Chapter 5.

Before attempting a careful study of Chapters 18 through 21, I find it desirable to cover all of Chapter 4 so that the students will have a fuller understanding of the concepts of mean and variance of random variables and sums of random variables.

An instructor, who feels that students who will spend only one semester on the study of statistics should have some acquaintance with t-tests, can replace Chapter 15 (nonparametric k-sample procedures) and Chapter 16 (rank correlation) with parts of Chapters 18 and 19 (one- and two-sample normal theory procedures, respectively).

ix

In Chapters 12 through 16 the following sections may be omitted without loss of continuity:

Chapter 12: Sections 87–90
Chapter 14: Sections 102 and 109   (Section 105 is needed only for an understanding of the Kruskal-Wallis test in Chapter 15)
Chapter 15: Sections 114, 116–119
Chapter 16: Section 125.

In a two-quarter course it should be possible to cover most of the material through Chapter 19. But again, an instructor may want to replace Chapter 14 with sections of Chapter 20 on the analysis of variance and/or Chapter 16 with sections of Chapter 21 on linear regression.

A few problems at the end of individual chapters are starred. These problems are somewhat theoretical in nature and in some cases require knowledge of Chapter 4.

I am grateful for permission to reproduce and/or use certain published tables. Acknowledgement is made at the appropriate place. I express my deep appreciation to Herman Chernoff, Ralph D'Agostino, and John Pratt who have read the entire manuscript and offered many helpful suggestions. I take full responsibility for any shortcomings. I am thankful to graduate students Michael Barthel and Teng-Shan Weng for checking computational details and to Miss Sandra Tamborello and my daughter Monica for typing the manuscript. Lastly, I am indebted to Houghton Mifflin Company for their assistance in bringing out this book.

Gottfried E. Noether

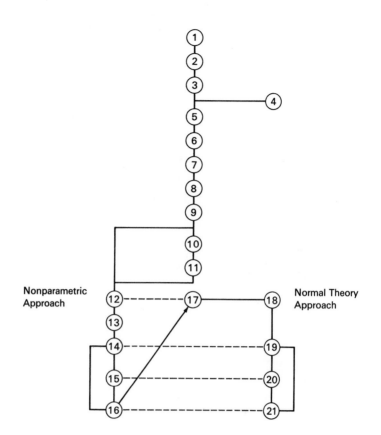

Nonparametric
Approach

Normal Theory
Approach

Contents

1

What Is Statistics?

"If experimentation is the Queen of the Sciences, surely statistical methods must be regarded as the Guardian of the Royal Virtue."
(From a letter to *Science* by Myron Tribus)

1 To most people the word "statistics" brings visions of endless columns of numbers, mysterious graphs, and frightening charts that show how the government is spending our tax money. At one time the word referred exclusively to numerical information required by governments for the conduct of state. Statisticians were people who collected masses of numerical information. Some statisticians still do, but many do not. They help in the conduct and interpretation of scientific experiments and professional investigations. Along with changes in the work of statisticians has gone a change in the meaning of the word "statistics" itself. "Statistics" may refer to numerical information, like football statistics or financial statistics. However the word also refers to a subject, a subject like mathematics or economics. When we use the word "statistics" in our discussions, we shall usually have this second meaning in mind. The few times when the word refers to numerical information as such, the fact will be clear from the context.

How, then, can we describe the field of statistics? A U.S. Civil Service Commission document says that "statistics is the science of the collection, classification, and measured evaluation of facts as a basis for inference. It is a body of techniques for acquiring accurate knowledge from incomplete information; a scientific system for the collection, organization, analysis, interpretation and presentation of information which can be stated in numerical form." I just hope that this definition does not keep anyone from wanting to study statistics!

Actually, the following statement taken from the book *The Nature of Statistics* by Wallis and Roberts is much more appropriate for our purposes: "Statistics is a body of methods for making wise decisions in the face of uncertainty." While this may not reflect as precise and all-inclusive a view of statistics as the earlier statement, it describes very succinctly that aspect of statistics that will be of greatest interest to us in this book, namely, how to use incomplete information to make decisions and draw valid conclusions.

The kind of decisions (or conclusions) that a statistician has in mind are decisions based on numerical information of one kind or another. Most of us are quite accustomed to making decisions of this sort. For example, after having noted on several occasions how much time it takes to get to the station or airport, we decide on how much time to allow in the future for such a trip. In general, experience coupled with common sense is quite enough to produce satisfactory results.

But there are many problems of a statistical nature for which common sense alone is unable to provide adequate answers. What are needed are more formal methods—methods of *statistical inference* as they are usually called—that have been developed and analyzed by mathematical statisticians using the calculus of probability. Starting with Chapter 6 we study some of the simpler such methods and the rationale behind them. In Chapters 2–5 we learn some basic probability. And during the remainder of this chapter we illustrate certain ideas that are basic for all of statistical inference. Of necessity this preliminary discussion will rely heavily on intuition.

The Taxi Problem

2   Suppose that you are waiting at a street corner for a taxi. Several drive by, but they are all occupied. You begin to wonder how many taxis there are in this city. Clearly there are not enough to go around. But how many? You start watching the numbers on the shields of the taxis going by:

$$405, 280, 73, 440, 179.$$

The next taxi stops and picks you up. You could simply ask your taxi driver how many taxis are registered in the city. But before you do so, you wonder whether on the basis of your observations you cannot make an educated guess. Indeed you can, as we shall see.

We represent our information graphically by marking the 5 observed numbers on a line as follows:

	73		179		280		405	440	
gaps:		72		105		100		124	34	?

The endpoint on the left corresponds, of course, to taxi #1. We should like to know the value of the endpoint on the right. How can we make a guess ? It is certainly reasonable to expect that the gap to the right of the largest observed number, 440 in our case, is similar in size to the observed gaps between the known numbers. So let us compute the average gap size and add it to the largest number. We then find

$$440 + \frac{72 + 105 + 100 + 124 + 34}{5} = 440 + \frac{435}{5}$$
$$= 440 + 87$$
$$= 527.$$

Using statistical terminology, we would say that our *estimate* of the total number of cabs is 527. Having arrived at an estimate, we are, of course, curious how good our estimate is. Our taxi driver informs us that there are 550 taxis in the city. Thus our estimate is in error by 23, less than 5 per cent, which is quite good considering that we observed only 5 out of 550 cabs.

By now, some students may be thinking that we are playing a rather silly game. Since we actually want to pursue the example further, it may be helpful to point out that this game, which goes by the name of *serial number analysis*, was most useful during World War II. There, German tanks took the place of taxi cabs. It was found at the end of the war that statistical estimates of German tank production were much more accurate than estimates based on more orthodox intelligence sources.

A little earlier we mentioned that we were going to illustrate certain basic ideas of statistical inference. Statistical inference is concerned with drawing useful conclusions from available data, usually called the *sample*, about a larger aggregate called the *population*. In our taxi example the sample consists of the taxis numbered 73, 179, 280, 405, and 440 that we happened to observe. The population consists of all taxis from 1 to 550. In this example the quantity that has the greatest interest is the number 550. Before the taxi driver told us, this number was unknown to us. But by looking at the 5 numbers in our sample, we were able to make a rather good guess. In other words, we found an *estimate* of the unknown quantity.

A different kind of problem arises if we have some preconceived idea of what the total number of taxis might be. Somebody might have assured us that there are at least 1000 taxis in the city. After waiting unsuccessfully for a taxi to pick us up and noting that the largest number in our sample is only 440, we may have some doubts about the claim that there are that many taxis around. We have a *statistical hypothesis* that we want to test. The hypothesis states that there are at least 1000 taxis in the city. But our sample information throws such great doubt on the correctness of the hypothesis that we prefer to believe otherwise. We may argue something like this. If it is really true that there are at least 1000 taxis around, in a sample of 5 there should ordinarily be one with a number greater than, say, 600. The fact that the largest number in our sample is only 440 makes the original hypothesis rather untenable. Presumably the true number of taxis in the city is considerably less than 1000. In Chapter 3 we get more definite evidence to this effect.

3   Estimation and tests of hypotheses are the two main problems of statistical inference that we are going to discuss in our course. There are many different aspects to testing and estimation. Another look at the taxi estimation problem suggests some of these.

It has perhaps occurred to some of you that there are other estimates of the total number of taxis than the one we have used. Another look at our graph suggests that the middle number in the sample, i.e., 280, is approximately halfway between the two endpoints. Since the upper endpoint represents the total number of cabs and the lower endpoint is 1, we can use this fact to find another estimate of the total number of cabs. However, before we look at details, let us introduce some symbols and define an important concept.

We denote the total number of cabs by the symbol T. Next, suppose that we have a set of numbers, like our taxi numbers. When we arrange these numbers according to size, say, from the smallest to the largest, the number in the middle is called the *median* of our set of numbers. In our group of 5 taxi numbers, the median is 280. In general, we denote the median of a set of numbers by the symbol M.

Our earlier remark can now be expressed as follows. The median M of our observed numbers is approximately halfway between 1 and T. Now, the exact value of the number halfway between 1 and T is $(1 + T)/2$, and, therefore, M is an estimate of $(1 + T)/2$. Accordingly, $2M - 1$ is an estimate of T. In our case,

$$2M - 1 = 2 \times 280 - 1 = 559,$$

so our second estimate of T is 559.

The fact that there is a second estimate raises the following problem. In practice, which estimate should we use, the first one or the second one, or possibly still a third one? Probably with a little thought, you can come up with still different ways of estimating the total number of taxis.

In our example, the second estimate is clearly better than the first estimate, since 559 is closer to the true value 550 than is 527. But of course, in general, there is no taxi driver around to tell us the true value after we have computed our different estimates. The Germans certainly did not tell the Allies how many tanks they had built during a given period of time. The Allies only found out for sure after the end of the war, when the information was purely academic. In most statistical problems, we never find out the true value.

How can we decide then which estimate to use, the first or the second? Before we give an answer to this question, let us look at some additional data, representing three more sets of 5 observations of taxi numbers. These numbers were taken from tables of *random numbers* rather than observing actual taxi-cabs. We will hear a great deal more about random numbers later on. At this moment all we have to know is that the results are comparable to an experiment in which three statisticians stand at three busy street corners in a city with 550 taxis numbered from 1 to 550, each writing down the numbers of the first five taxis that happen to pass by. The numbers in our original sample as well as those in the additional three samples are listed in Table 1.1. Easy com-

TABLE 1.1						
	Sample 1	405	280	73	440	179
	Sample 2	72	132	189	314	290
	Sample 3	485	65	108	382	298
	Sample 4	450	485	56	383	399

putations produce the estimates in Table 1.2, where the first

TABLE 1.2	gap estimate		median estimate	
Sample 1	527	(23)	559	(9)
Sample 2	376	(174)	377	(173)
Sample 3	581	(31)	595	(45)
Sample 4	581	(31)	797	(247)

estimate is called the gap estimate and the second estimate is called the median estimate, and where the numbers in parentheses indicate by how much each estimate is in error.

One thing is clear from these computations. Neither estimate is consistently better than the other. In two samples the gap estimate happens to be closer to the true value; in one case the median estimate is closer. And in one case, the two estimates practically coincide. On the basis of the available information a clear-cut decision as to which estimate is better is not possible. However, by continuing the experiment we would eventually find that on the average, the gap estimate is somewhat closer to the true value than the median estimate. Some indication of this can be found in our samples. On the average the first estimate deviates less from the true value than the second estimate. While the average error of the median estimate is 118.5, that of the gap estimate is only 64.8. This is where the mathematical statistician comes in. With the help of the theory of probability a mathematical statistician can not only determine which of two estimates is better, but can also determine how much better one estimate is than the other. However such investigations are beyond the scope of our course. In general we have to take the word of a mathematical statistician that a recommended procedure has desirable properties.

There is a more convenient way for computing the gap estimate.

Let
$$n = \text{number of observations in sample}$$
$$L = \text{largest observation in sample.}$$

Then we have:

1.3
$$\text{gap estimate} = L + \text{average gap}$$
$$= L + \frac{L - n}{n} = \frac{nL + L - n}{n}$$
$$= \frac{n + 1}{n} L - 1.$$

For example, for the first sample we find
$$\tfrac{6}{5} \times 440 - 1 = 6 \times 88 - 1 = 527$$
as before.

Assumptions

4   Of necessity the previous discussion has been rather intuitive. However, in view of later developments one additional point should be mentioned. Underlying our intuitive approach are various *assumptions* without which the suggested solutions would have very doubtful value. For example, we clearly assumed that taxis in the given city were numbered consecutively from 1 to T. This seems like a reasonable assumption but need not necessarily be true. For instance, in the corresponding problem of estimating German tank production, it soon became evident that German tanks were not numbered in any one consecutive sequence, complicating the estimation problem considerably. Nearer to home, the author conducted the following experiment. In one of the Boston suburbs, he noted down the numbers of the first 5 taxis that crossed a certain main intersection and showed evidence of being registered locally, with the following results:

35, 18, 38, 43, 23

Our gap estimate becomes $\tfrac{6}{5} \times 43 - 1 = 50.6$, so that we would estimate T as either 50 or 51. Inquiry revealed that there were actually only 40 taxis registered in town. The estimate was in error by over 25%. The reason is easy to see. Since the largest observed number was actually 43, not all numbers could have been in use. Indeed, even a cursory inspection of the actually observed numbers reveals an absence of low numbers. Apparently all or most taxis with lower numbers have been taken out of use. (See Problem 9 of Chapter 3.)

A second assumption which is more difficult to check is that each taxi is equally likely to pass the corner where we are waiting to be picked up. This assumption would not be appropriate if, for example, taxis 1–100 are used during daylight hours and taxis 101–200 at night.

The important point is to realize that statistical conclusions are not absolute. They are valid only within a given framework. The more realistic a framework we have chosen, the more reliable our conclusions will be. A given statistical method may be very appropriate within one framework, but quite inadequate within another. An important part of a statistician's work is to select a reasonable framework—or probability model, as we shall usually say—and then use methods of statistical analysis that are appropriate for this model. It should be clear that a statistical technique which is based on assumptions of a relatively general nature is often preferable to a more elaborate technique assuming a rather restrictive framework. As is pointed out more clearly later on, the nonparametric methods discussed in Chapters 12–16 have very general validity, while the corresponding methods in Chapters 18–21 assume a more stringent framework.

Medians

5   The median estimate discussed above requires the determination of the median M of a set of numbers. Since the median is used repeatedly in subsequent chapters in connection with other statistical problems, we close the chapter with a discussion of its computation.

In our examples, the computation of the median presented no problems. Among 5 observations, all different, the median is simply the observation "in the middle," i.e., the third largest (or smallest) among all 5 observations. However this description is not sufficiently precise to determine the median of an arbitrary set of numbers. We need a general definition of the median and a simple procedure for determining it in practice. The median M of a set of n numbers has the following property: there are (as nearly as possible) as many numbers that are smaller than M as there are numbers that are greater than M. In particular if $n = 2k + 1$ is odd, the median equals the $(k + 1)$st largest (or smallest) number in the set. If $n = 2k$ is even, any number between the kth smallest and the kth largest numbers satisfies our condition. For definiteness, we then usually choose the number halfway between the kth smallest and kth largest numbers as the median.

EXAMPLE   The median of 405, 280, 73, 440, 179 is 280.
1.4   The median of 405, 280, 73, 440 is any number between 280 and 405. For definiteness we choose $(280 + 405)/2 = 342.5$.
The median of 405, 280, 440, 179, 405 is 405. (In this example the third and fourth largest numbers are both equal to 405. Since by definition the median M equals the third largest, $M = 405$. In this example two numbers are smaller than M; only one number is greater than M.)
The median of 405, 280, 405, 405 is 405. (Both the second and third smallest numbers equal 405.)

As long as *n* is not too large, a convenient method for determining the median *M* is as follows. We cross out the largest and the smallest numbers and continue crossing out numbers in pairs until either one or two numbers are left. At each step the largest and smallest numbers that have not been previously eliminated are crossed out. If at the last step a single number is left, this number is *M*. If two numbers are left, *M* is chosen as the number halfway between the two numbers. Any number that occurs more than once is used in this crossing-out process as often as it occurs.

EXAMPLE
1.5

For the set 405, 280, 73, 440, 179, 405, we start by crossing out 440 and 73. This leaves 405, 280, 179, 405. Now we cross out 405 and 179, leaving 280 and 405. Thus $M = (280 + 405)/2 = 342.5$.

PROBLEMS

1 Find the median for each of the following sets of numbers:
   a. 27, 13, 14;
   b. 35, 49, 68, 50;
   c. 25, 25, 25;
   d. 46, 26, 39, 45, 26, 11, 51;
   e. the 20 taxi numbers in Table 1.1.

2 Check the gap and median estimates for samples 2 through 4 in Table 1.2.

3 Find the gap and median estimates for the following two sets of taxi numbers:
   a. 309, 769, 78, 61, 277, 188;
   b. 209, 181, 595, 799, 694, 334, 595.
   Remark: For the gap estimate, use Formula 1.3 rather than the original gap procedure.

4 Consider the four samples of size 5 in Table 1.1 as one sample of size 20. Find the gap and median estimates.

5 Get a sample of 10 taxi numbers for your city. Estimate *T* using the gap estimate. Try to find out the true value of *T* for comparison. If your estimate is in error by, say, more than 20%, try to find reasons.

6 Suggest one or two other methods for estimating the number *T*. Compute the corresponding estimates for the usual four samples. Do they seem to be better or worse than the gap estimate, the median estimate?

7 Assume that the fifth taxi number in sample 2 of Table 1.1 is 390 (rather than 290). Recompute the gap and median estimates. Why does this example show that the median estimate may produce rather undesirable estimates?

2

The Meaning of
Probability

The Frequency Interpretation

6 In Chapter 1 we saw that probability considerations form the basis of statistical inference. In this and the next chapter, we discuss some simple aspects of probability.

We start our discussion by introducing some notation. Probabilities are associated with events, like the event that it rains. We usually use letters like A, B, and so on, to denote events. For the probability of the event A we write $P(A)$ and read "the probability of A" or more briefly, "P of A." As an example, in books on card games we read that when we deal five cards from a well-shuffled deck of playing cards, the probability is $\frac{198}{4165}$ that the five cards contain "two pairs" (such as 2 kings, 2 tens and an ace). If we denote the event "two pairs" by the letter T, we would write $P(T) = \frac{198}{4165}$. Probabilities like this one are not difficult to compute, provided we know some basic principles of what the mathematician calls combinatorial analysis, but right now we are more interested in knowing what such a probability means. Actually this particular probability is too elaborate for such a discussion. On the other hand, we note that $\frac{198}{4165}$ does not differ very much from $\frac{200}{4000} = \frac{1}{20}$, so instead, we consider what we mean when we say that a certain event has probability $\frac{1}{20}$.

When we deal 5 cards from a deck of cards they either contain two pairs or they do not. There are no other possibilities. Where does the $\frac{1}{20}$ come in? The answer to this question can be found in certain empirical facts that were noted by gamblers hundreds of years ago. Consider any game of chance. If the game is played honestly, we cannot predict the outcome of a single game, or *trial*, as we shall often say. But something rather remarkable happens when we play the same game over and over again. If we look only at the outcomes of successive trials, no pattern emerges. However, the relative frequency with which any given outcome occurs shows greater and greater regularity and eventually becomes nearly constant. When we talk of the probability of an event, we have this limiting value of the relative frequency in mind.

7   Table 2.1 illustrates this phenomenon. The data represent a partial record of what happened when a "coin" was tossed 5000 times. We have recorded how often heads occurred during the first 10 tosses, the first 20 tosses, and so on. By dividing the number of heads by the number of trials, we obtain the relative frequency of heads.

*TABLE
2.1*

Results of 5000 Coin Tosses

number of trials	number of heads	relative frequency
10	7	.700
20	11	.550
40	17	.425
60	24	.400
80	34	.425
100	47	.470
200	92	.460
400	204	.510
600	305	.508
800	404	.505
1000	492	.492
2000	1010	.505
3000	1530	.510
4000	2030	.508
5000	2517	.503

The sequence of relative frequencies holds the greatest interest for us. During the early stages of the experiment the relative frequencies show considerable variability, but after hundreds and thousands of trials they become more nearly constant. Presumably if the experiment were continued, there would be less and less fluctuation in the relative frequency, suggesting eventual convergence to a constant which would then be called the probability of heads.

8   The interpretation of probability that we have just discussed is called the frequency interpretation of probability, for obvious reasons. For our purposes the frequency interpretation is extremely useful. However, other interpretations of probability are also possible. There are situations where the notion of unlimited repetition of an experiment as required by the frequency interpretation is, to say the least, very artificial. For example, people talk of the probability that the Boston Red Sox will win the pennant this coming year. Here we are interested in one particular pennant race. It is possible to argue that the same is true of the toss of a coin. But we can easily think of the toss of a coin as one of a long sequence of similar tosses. The same cannot be said of the American League pennant race with its ever-changing conditions. We then often speak of *personal* or *subjective* probabilities. Such probabilities give numerical expression to a person's *intensity of belief*, that is, his conceivable willingness to bet a certain amount of money on the occurrence or nonoccurrence of a given event.

From the purely mathematical point of view, personal probabilities can be dealt with in the same way as probabilities that have a frequency interpretation. They can be and often are made the basis of decisions in the face of uncertainty. However, when we talk of probabilities, we shall usually have a frequency interpretation in mind.

The eventual constancy of the relative frequency of events associated with repetitive experiments is not a property of games of chance alone. While the theory of probability obtained its start from problems raised by gamblers in the 17th century, its spectacular growth, particularly during the last thirty or forty years, is due to the realization that the methods that give answers to questions posed by gamblers also answer questions of scientists, engineers, and businessmen. Indeed it is difficult to imagine a field of human endeavor in which probability is not used in one way or another. One of its most important applications is in the field of statistics.

We repeat what we have learned about the frequency interpretation of a given probability. When we say that the probability of a given event is, for example, $\frac{1}{20}$, we imply that the relative frequency with which the event would occur in a long sequence of trials eventually stabilizes at the value $\frac{1}{20}$. It is customary to express this result by saying that we expect the event to happen about once in every twenty trials, or about five times in one hundred trials. There is no harm in such a statement if we remember its correct meaning in terms of what happens in the long run and do not interpret it literally. It certainly does not mean, as many people seem to think, that if an event with probability $\frac{1}{20}$ has not occurred in 19 successive trials, it is bound to occur on the twentieth trial. On the twentieth trial the event still has only one chance in twenty of occurring.

Random Numbers

9   We conclude this chapter with a brief discussion of *random digits,* or *random numbers* as they are often called. Some students may have wondered about the use of quotation marks in the description of the coin tossing experiment. The explanation is that no coin was used in collecting the data. (Real coin tosses take a great deal of time and require careful attention to keeping conditions unchanged from one toss to another.) Instead, the basic information came from a table of random digits. We have mentioned a table of random digits once before in connection with the taxicab problem. What then is a table of random digits?

Let us consider the following experiment. We have 10 identical pingpong balls on which we have written the digits 0, 1, . . . , 9. After mixing the balls very thoroughly in a box, we select one of the balls without looking, note down the digit written on the ball, and then put it back in the box. We then repeat the whole process of mixing, selecting, noting down a digit, and returning the ball to the box again and again. The result will look something like this:

$$40582 \quad 00725 \quad 69011 \quad 25976 \quad \ldots\ldots,$$

where the digits have been written in groups of five for easier reading. A table of random digits is simply a more extensive listing, except that such a listing is usually produced by an electronic computer instead of an actual experiment like ours.

In Chapter 3 we discuss random digits in greater detail. Now we simply observe that random digits have the property that each of the ten digits 0 through 9 has probability $\frac{1}{10}$ of occurring in any position in a random number table.

Random digits can be used to simulate a coin tossing experiment like the one we described earlier. Each digit in a sequence of random digits may be assumed to furnish the result of a coin toss by letting, for example, even digits correspond to heads and odd digits, to tails. In this way the first ten digits in our listing produce seven heads and three tails; all twenty digits produce eleven heads and nine tails. (These are the first twenty "tosses" in Table 2.1.)

Random digits play an important role in many statistical investigations. Many high-speed computers have built-in provisions for producing random digits as needed. There are also available printed tables, the largest of which contains 1 000 000 digits. The 5000 digits that formed the basis of our coin tossing experiment were taken from this table. Table J at the end of the book is a short table of random numbers.

PROBLEMS

1 Roll a die 120 times and observe how often it falls 1, 2, . . . , 6. Are the results in agreement with what you would have expected in such an experiment? (Keep your data for later use.)

2 Imagine an experiment in which you roll two dice 180 times and observe how often the two dice add up to 2, 3, . . . , 12. Invent appropriate data for such an experiment. (Be sure to do this problem before doing Problem 3. Keep the results.)

3 Actually perform the experiment described in the preceding problem. Are the results similar to those for Problem 2? (Keep the data.)

4 Imagine the same kind of experiment as in Problem 1, but involving 6000 rolls. Invent appropriate data and keep them for later use.

3

Probability:
Some Basic Results

Union and Intersection of Events

10   Probabilities are computed in many different ways, depending on the circumstances under which they arise. Actually the practicing statistician rarely has to compute probabilities on his own. Nearly all the probabilities that a statistician needs in his daily work have already been computed for him and are read from appropriate tables. Even so, for a better understanding every statistician should be familiar with certain basic principles and results.

In Chapter 2 we saw the close relationship between relative frequencies and probabilities. Since relative frequencies take values between 0 and 1, the same is true of probabilities. If A is any event, the probability of the event A has a value between 0 and 1, in symbols,

$$0 \leq P(A) \leq 1.$$

An event that never occurs has probability 0. An event that always occurs has probability 1.

11   For our further discussion we consider a ficticious college whose student enrollment is described in Table 3.1.

TABLE
3.1

College Enrollment

	men	women	all students
freshmen	200	100	300
sophomores	175	75	250
juniors	175	75	250
seniors	150	50	200
all classes	700	300	1000

One of the students is to represent the college at a national convention. When everybody volunteers, it is decided to select a student by lot. Each student's name is written on a separate slip of paper. The 1000 slips are then thoroughly mixed in a large bowl, and, as soon as the college president returns from a speaking tour to alumni chapters, he is going to select one of the slips to determine the name of the lucky winner. We want to determine what the probability is that the selected student will be (*i*) a woman, (*ii*) an upperclassman, (*iii*) a male freshman.

First of all we introduce symbols to stand for events of interest:

> F, for freshman, or more exactly, for the event that the name selected by the college president is that of a freshman
>
> J, for junior
>
> S, for senior
>
> U, for upperclassman
>
> M, for man
>
> W, for woman

We then write $P(W)$ for the first probability we are interested in. Similarly we can write $P(U)$ for the second probability. But then we notice that the event U is actually a combination of events J and S. A student is an upperclassman if he is either a junior *or* a senior. Thus $P(U)$ is the same as $P(J \text{ or } S)$ and it is reasonable to ask whether this probability can be expressed in terms of $P(J)$ and $P(S)$. As far as the third probability is concerned, we have not defined a special symbol for male freshmen. But since a student must be both male *and* a freshman in order to be classified as a male freshman, we write symbolically $P(M \text{ and } F)$. Let us then find $P(W)$, $P(J \text{ or } S)$, and $P(M \text{ and } F)$.

The experiment of selecting one name from among 1000 names reminds us of an earlier experiment, the selection of one pingpong ball from among 10 (representing the digits 0, 1, ... , 9). In that example we associated probability $\frac{1}{10}$ with any particular digit. In the present case we associate probability $\frac{1}{1000}$ with any particular name. Further, since there are 300 slips of paper corresponding to women students, we set $P(W) = \frac{300}{1000}$ or .30. Similarly $P(U) = P(J \text{ or } S) = \frac{450}{1000}$ or .45, since there are 450 upperclassmen at the college. Finally $P(M \text{ and } F) = \frac{200}{1000}$ or .20, since there are 200 male freshmen at the college.

While we have found the probabilities $P(J \text{ or } S)$ and $P(M \text{ and } F)$, we have not yet answered the more interesting question of how these probabilities are related to $P(J)$, $P(S)$, $P(M)$, and $P(F)$. We easily find $P(J) = \frac{250}{1000}$ or .25; $P(S) = \frac{200}{1000}$ or .20; $P(M) = \frac{700}{1000}$ or .70; $P(F) = \frac{300}{1000}$ or .30. Thus $P(J \text{ or } S) = P(J) + P(S)$. But there does not seem to be any easily recognizable relationship between $P(M \text{ and } F)$, on the one hand, and $P(M)$ and $P(F)$, on the other hand. To bring out an additional complication, we consider the following problem. On campus there is a social club whose membership, in view of the shortage of women students, consists of all the female students but only the male students who are upperclassmen. What is the probability that the student selected by lot belongs to this social club? Since there are $300 + 175 + 150 = 625$ members, the answer is $\frac{625}{1000}$ or .625. But let us express this probability by means of our customary symbols. Since a student who belongs to the social club must be either an upperclassman or female (or both), we have $P(U \text{ or } W) = .625$. On the other hand, $P(U) + P(W) = .45 + .30 = .75$. This time the two are not equal. How can we resolve our difficulties? Some general terminology is helpful for our discussion.

Let A and B be any two events associated with a chance experiment. We define two new events denoted by "A and B" and "A or B", respectively. The event "A and B" is often called the *intersection* of events A and B, while the event "A or B" is called the *union*. The intersection requires that *both* events A and B are satisfied. The union requires that one or the other event is satisfied. (This definition does not preclude the possibility that both events are satisfied. Thus we can also say that the union requires that *at least one* of the two events is satisfied.) We want to find formulas for $P(A \text{ and } B)$ and $P(A \text{ or } B)$.

The Addition Theorem of Probability

12 We start with the formula for the union of two events. Returning to our college example, we remember that $P(J \text{ or } S)$ equals $P(J) + P(S)$, but $P(U \text{ or } W)$ does not equal $P(U) + P(W)$. The reason for these two different results is not difficult to discover. Events J and S are *exclusive:* a student who is a junior cannot at the same time also be a senior. Events U and W are not exclusive: a female student may well be an upperclassman. When computing the value of $P(U) + P(W)$, we count 125 female upperclassmen twice, once when computing $P(U)$ and a second time when computing $P(W)$. In order to obtain a correct expression for $P(U \text{ or } W)$, we have to adjust for this double count by subtracting $P(U \text{ and } W) = \frac{125}{1000}$, the probability that a student is an upperclassman and female. We then obtain the correct result

$$P(U \text{ or } W) = \tfrac{450}{1000} + \tfrac{300}{1000} - \tfrac{125}{1000} = \tfrac{625}{1000}.$$

More generally we have the following formula for the probability of the union "A or B" of two events A and B:

3.2 $$P(A \text{ or } B) = P(A) + P(B) - P(A \text{ and } B).$$

Formula 3.2 is often called the *Addition Theorem of Probability*. If events A and B are exclusive, $P(A \text{ and } B)$ is zero and (3.2) becomes:

3.3 $$P(A \text{ or } B) = P(A) + P(B),$$

if A and B are exclusive events.

Independent Events

13 We now turn to the probability for the intersection of two events. For events M and F in the college example we have already seen that there does not seem to exist an obvious relationship between $P(M \text{ and } F)$ on the one hand and $P(M)$ and $P(F)$ on the other hand. But let us look at a different example.

Suppose we select one card from a well-shuffled deck of playing cards. Let Q be the event that the selected card is a queen, and R the event that the card belongs to a red suit. Following our earlier examples we easily find $P(Q) = \frac{4}{52} = \frac{1}{13}$, $P(R) = \frac{26}{52} = \frac{1}{2}$, and $P(Q \text{ and } R) = \frac{2}{52} = \frac{1}{26}$, so that in this example $P(Q \text{ and } R) = P(Q)P(R)$. We say that events Q and R are *independent*. More generally, two events A and B are said to be independent if

3.4
$$P(A \text{ and } B) = P(A)P(B).$$

The concept of independence of two events occupies an important place in probability and statistics. If the probability setup of an experiment is completely known, we can find out whether or not two events A and B associated with the experiment are independent by checking condition (3.4). But in statistical applications we often use (3.4) in a different way. The *experimental* situation may be such that we would expect two events A and B associated with the experiment to be independent. We then simply go ahead and compute the probability $P(A \text{ and } B)$ as the product $P(A)P(B)$. Thus it is important to know under what conditions it is realistic to assume that two events are independent. Let us see what the frequency interpretation of probability suggests.

We write $\#(A)$ to denote the number of occurrences of the event A in n trials. More generally, $\#(\ )$ denotes the number of occurrences of the event indicated within parentheses. According to the frequency interpretation, $P(A)$ is approximately equal to the relative frequency of the event A in a large number of trials. We can write this as

$$P(A) \doteq \frac{\#(A)}{n},$$

where $\doteq$ denotes approximate equality. By definition two events A and B are independent if

$$P(A \text{ and } B) = P(A)P(B).$$

Thus if events A and B are independent we must have

$$\frac{\#(A \text{ and } B)}{n} \doteq \frac{\#(A)}{n} \frac{\#(B)}{n}.$$

But this can be written as

$$\frac{\#(B)}{n} \doteq \frac{\#(A \text{ and } B)}{\#(A)}, \quad (\text{provided } \#(A) > 0).$$

The meaning of the expression on the left is clear. It is the relative frequency with which B occurs among all n trials. But the expression on the right can also be considered as a relative frequency: in the denominator there is the number of occurrences of A; in the numerator, the number of times both A and B occur jointly. The ratio of the two expressions gives the relative frequency with which B occurs among the trials that result in the occurrence of A. It follows that if A and B are independent in the probabilistic sense, the frequency with which B occurs among the trials at which A also occurs is approximately the same as the frequency with which B occurs in all n trials. Under such circumstances the knowledge that A has or has not occurred is of no help when we wonder whether B has or has not occurred as the result of the same experiment. If we interchange A and B in this discussion, the reverse also follows. In practical applications then, it is realistic to assume that two events A and B are independent when the experimental situation is such that the occurrence or non-occurrence of one of the events does not suggest any change in our evaluation of the probability of the other event.

Let us return for a moment to our earlier examples. Since every suit in a deck of playing cards contains the same cards, namely ace, king, and so on, knowledge of the suit of a card does not reveal anything about whether or not the card is a queen. We are just as likely to draw a queen from among red (or black) cards as from among all 52 cards. Knowledge of the suit of a card does not reveal anything about its value.

The situation is different for events F and M in the college example. The proportion of male students among freshmen $(\frac{200}{300} = .667)$ is somewhat lower than the corresponding proportion among non-freshmen $(\frac{500}{700} = .714)$. Knowing that a student is a freshman reduces the chances that he is male.

Conditional Probabilities

14 Let us consider again events U and W in the college example. We have already observed that events U and W are not exclusive. Are they independent? Since $P(U$ and $W) = .125$ and $P(U) \times P(W) = .450 \times .300 = .135$, the answer is no. But let us be more specific. The probability of selecting a woman student among upperclassmen is different from that of selecting a woman student among freshmen and sophomores. In particular, there are 450 upperclassmen, of whom 125 are women. If one of the upperclassmen is selected by lot, the probability that this upperclassman is a woman is $\frac{125}{450}$ or .278. Since there are 550 freshmen and sophomores, of whom 175 are women, the corresponding probability of selecting a woman from among freshmen and sophomores is $\frac{175}{550}$ or .318. It is important that we distinguish these two probabilities from each other and from the earlier probability $P(W) = .300$. It is customary to refer to the two new probabilities as *conditional* probabilities, since they are computed subject to supplementary conditions. Rather than draw lots among all college students as we do in connection with probability $P(W)$, we now draw lots among only part of the college population. We draw attention to this fact by writing $P_U(W) = .278$ and $P_L(W) = .318$, where L denotes the event "the student is either a freshman or a sophomore."

More generally, if A and B are two arbitrary events, we write $P_A(B)$ for the probability of the event B computed under the assumption that event A is satisfied. As for the corresponding frequencies, we have

$$P_A(B) = \frac{P(A \text{ and } B)}{P(A)},$$

which can also be written in the form

$$P(A \text{ and } B) = P(A)P_A(B).$$

This formula allows us to compute the probability of the event "A and B" even if A and B are not independent. For example,

$$P(U \text{ and } W) = P(U)P_U(W)$$
$$= \tfrac{450}{1000}\,\tfrac{125}{450} = .125$$

as before.

If events A and B are independent,

$$P_A(B) = \frac{P(A \text{ and } B)}{P(A)} = \frac{P(A)P(B)}{P(A)} = P(B),$$

showing again that for independent events the occurrence or nonoccurrence of one of the events has no bearing on the probability of the other event.

Another Look at Random Numbers

15 We have seen that in a table of random numbers, all digits from 0 to 9 have the same probability $\frac{1}{10}$ of appearing in a given position. But random digits can also be read in groups of 2, or groups of 3, and so on. We can then ask ourselves what the probability is that a sequence of 2 random digits forms a number like 53. We have

$$P(53) = P(\text{first digit} = 5 \text{ and second digit} = 3).$$

But what is the probability on the right? Our method of producing random digits certainly implies that successive digits are independent. The reason for mixing our pingpong balls very thoroughly before drawing out another ball is to insure independence of successive draws. Thus $P(53) = P(5)P(3) = \tfrac{1}{10} \times \tfrac{1}{10} = \tfrac{1}{100}$, and a corresponding result holds for groups of 3 or more random digits. As an example, $P(539) = P(5)P(3)P(9) = \tfrac{1}{1000}$.

When looking at random number tables, people are often surprised to find double or triple or even quadruple entries like 33 or 555 or 8888. Somehow the human mind seems to feel that such sequences have no place in a random scheme. But let us look at the question from the probabilistic point of view.

We have seen that any two-digit random numbers like 53 or 87 or 22 have the same probability $\frac{1}{100}$. Since there are 10 double numbers, 00, 11, and so on, the probability of having a two-digit random number with 2 identical digits is $\frac{10}{100} = \frac{1}{10}$. Thus, on the average, one of every 10 pairs of random digits should be a double number. There is an even simpler and more instructive way of deriving this probability by remembering how we obtained random numbers with the help of pingpong balls. We are simply asking for the probability that on the second try we draw the same ball as on the first try. Since there are 10 balls from which to choose, the probability is $\frac{1}{10}$. A similar argument shows that the probability of identical triplets of digits is $\frac{1}{100}$ and the probability of identical quadruplets is $\frac{1}{1000}$.

16 A little earlier we remarked that our idea of randomness does not always agree with randomness as produced by mechanical methods. Students in an elementary statistics course were asked to write down what they considered to be a sequence of random digits. More exactly, they were asked "to mentally draw a sequence of pingpong balls." To simplify matters, only three balls labeled 1, 2, and 3 were to be used in this mental game. The record of 900 two-digit sequences produced in this way is given in Table 3.5.

TABLE 3.5

Frequency of Guessed Digit Pairs

		second digit			
		1	2	3	totals
	1	62	144	119	325
first digit	2	114	54	127	295
	3	106	116	58	280
totals		282	314	304	900

We should like to find out whether real and imagined draws of pingpong balls produce similar results. For real draws the probability is $\frac{1}{3}$ of obtaining each of the digits 1, 2, and 3. Thus in 900 draws we would expect to find something like 300 1's, 2's, and 3's each. If we look at the column on the right, we see that the students participating in the experiment wrote down 325 1's, 295 2's, and 280 3's as their first digit. Most of us will agree that these results look perfectly reasonable. We shall get statistical evidence to this effect in a later chapter. The same is true of the frequencies for the second digit given in the bottom row of our table.

But let us now look at pairs of numbers. There are 9 possible pairs: 11, 12, 13, 21, 22, 23, 31, 32, 33. For real draws each pair has the same probability $\frac{1}{3} \times \frac{1}{3} = \frac{1}{9}$. Thus in an experiment involving real pingpong balls, we would expect to find about 100 observations in each one of the 9 cells. When we look at the experimental data, it is fairly obvious that there are not enough entries down the diagonal from the upper left to the lower right, and too many entries off the diagonal. Students evidently were hesitant to write down pairs of identical numbers in their random sequences. For the sake of comparison, the results of an experiment using real pingpong balls is given in Table 3.6. The totals

TABLE 3.6

Frequency of Random Digit Pairs

		second digit			
		1	2	3	totals
first digit	1	94	89	118	301
	2	104	90	87	281
	3	108	107	103	318
	totals	306	286	308	900

are very similar to those in Table 3.5. But the entries in the body of the table are completely different. This time there is no particular difference between the entries down the diagonal and those off the diagonal. This, of course, is as it should be for random pairs of digits.

We shall return to this example on several later occasions after we have discussed appropriate statistical methods for analyzing this type of data. Such a statistical analysis will not cause us to change our present impressions. It will simply help to put them on firmer foundations.

The reason for discussing the example now is to show the difference between independence and dependence. Successive digits produced by real draws of pingpong balls are independent (provided sufficient mixing of balls between successive draws takes place). The same is not true of mental draws. The human mind remembers earlier choices. This remembrance, though possibly unintentional, introduces an element of dependence.

We can draw an important conclusion from the experiment. The human mind does not seem to be able to imitate real random behavior. In Chapter 1 we said that statistical inference is concerned with generalizations based on sample information. However, statistical procedures provide a basis for valid generalizations only if the sample information has been obtained according to known laws of the theory of probability. It follows that in collecting sample information it is advisable to use mechanical means like random number tables or similar devices to insure proper random selection. Experiments like the one we have just discussed show that human beings cannot be trusted to obtain truly random samples if the actual selecting is left to them.

17   ***Application to Sampling***   In conclusion then, we want to indicate how random numbers can often be used to obtain random samples. Consider the following rather common problem. We want to survey student attitude at a given university on some topic such as smoking marijuana. By one method or another we decide to include 225 students in our sample. How are we going to select them? We might simply take the first 225 students entering the dining room for breakfast one morning. But what about the students who eat at home or do not feel like eating breakfast that morning? It is quite possible that students living at home feel differently about smoking marijuana than students living in dormitories.

We want a method of selection that chooses students one at a time and with equal probability. The problem is quite simple if there exists a central file of all students at the university. We can then assign each student an identity number. Indeed, at many universities students already have identification numbers. For simplicity let us assume that these numbers go consecutively from 1 to, say, 7629. Since the highest possible number has four digits, we take a table of random digits and consider them in blocks of four. Suppose this is the way our table looks:

$$\ldots\ldots\ 18048\ \ 25400\ \ 76364\ \ldots\ldots$$

The first student to be included in our sample is the one with identity number 1804. Since no student has the number 8254, we simply skip this number. The next student has number 0076, or simply 76. We continue in this way until we have obtained 225 identity numbers. Of course, now the real work begins. We have to get the names and addresses of the 225 students and contact them. Other methods of selection may be simpler and faster. But this method has the all-important property of true randomness.

18   When using a random number table to select a random sample it may happen that the same number appears more than once. Two possibilities are then open to us. We can record a student's opinion as often as his identity number appears in the table of random numbers. The resulting sampling procedure is known as sampling *with replacement*. Alternatively we can decide to ignore a number after its first appearance exactly as we ignore a number that does not correspond to a member of our student population. The resulting sampling procedure is known as sampling *without replacement*. Before taking a sample we have to decide which of the two sampling procedures is appropriate under the circumstances.

EXAMPLE 3.7   A new student organization has ten members, seven men and three women. They decide to leave the election of a president and secretary to chance. What is the probability that both offices will be filled by women?

If the students do not mind having both offices filled by one and the same person, they can use sampling with replacement. When sampling with replacement, successive choices are independent and the desired probability is simply

$$\tfrac{3}{10} \times \tfrac{3}{10} = \tfrac{9}{100} = .090.$$

If the students want to be sure that the offices of president and secretary are filled by two different persons, sampling with replacement is inappropriate. When sampling without replacement, computations are slightly more complicated. We now need the formula $P(A \text{ and } B) = P(A)P_A(B)$ for the simultaneous occurrence of two events that are not independent. In the present example, A is the event that the student elected to the office of president is a woman, while B is the event that the student elected to the office of secretary is a woman. We have $P(A) = \tfrac{3}{10}$ and $P_A(B) = \tfrac{2}{9}$, so that

$$P(\text{both offices are filled by women}) = \tfrac{3}{10} \times \tfrac{2}{9} = \tfrac{6}{90} = .067$$

when sampling without replacement.

The Taxi Problem Revisited

19   In Chapter 1 we mentioned briefly the need for setting up an appropriate framework—or probability model, as we shall say from now on—for solving a statistical problem. The two assumptions we mentioned in connection with the taxi problem, consecutive numbering from 1 to T and equal likelihood for any one taxi to pass the spot where we are standing, may be translated into more mathematical language by saying that the 5 taxi numbers in Chapter 1 constitute a random sample of size 5 from a population of tags numbered from 1 to 550. As we have seen, such a sample can be obtained by selecting 5 three-digit numbers from a random number table, ignoring combinations 551–999 and 000. Indeed the samples in Chapter 1 were obtained in exactly this fashion.

We want to investigate more precisely why in Chapter 1 we decided to reject the hypothesis that there are at least 1000 taxis in the given city. For our investigation we assume that the random number model is appropriate for analyzing taxi data. Let us assume for the moment that there are exactly 1000 taxis. In our mathematical model, these taxis are represented by the three-digit numbers 001, 002, . . . , 999, 000 (in place of 1000). The largest observed taxi number in our sample was 440. Let us then compute the probability that each of 5 three-digit random numbers selected from all three-digit numbers is 440 or smaller. The probability that any one such number is smaller than or equal to 440 is $\frac{440}{1000} = .44$. Our model implies independence from one selection to another. Therefore the probability that all 5 numbers are smaller than or equal to 440 is $(.44)^5 = .016$. Thus according to the frequency interpretation of probability, we should expect to observe such an event only about 1.6 times in 100 trials if, indeed, there are 1000 taxis in the city, and even less frequently if there are more than 1000 taxis. It is of course always possible that we actually did observe an event having probability .016 or less. On the other hand, a much more plausible explanation is that the number 1000 in the probability $\frac{440}{1000}$ is incorrect. As a consequence we decide that the claim that there are at least 1000 taxis at our disposal is very doubtful. In statistical terminology, we decide to reject the hypothesis being tested.

20 It is instructive to pursue the problem somewhat further. We have already rejected the hypothesis that there are at least 1000 taxis. But what about claims of at least 900, or 800, or perhaps only 700? Corresponding hypotheses can be tested in very much the same way. If $T = 900$, the probability that any one taxi number is smaller than or equal to 440 is $\frac{440}{900} = .49$, and the probability that all 5 taxi numbers are smaller than or equal to 440 is $(.49)^5 = .028$, still rather small. Very likely, there are not even 900 taxis around. When we come to $T = 800$, we find $\frac{440}{800} = .55$ and $(.55)^5 = .050 = \frac{1}{20}$. As we shall see in later chapters, statisticians often use $\frac{1}{20}$ as a dividing point between "small" probabilities suggesting rejection of a hypothesis and sufficiently large probabilities that have no such implication. If we follow this practice, we would then claim that T is smaller than or equal to 800, or in symbols, $T \leq 800$. We have then found a "reasonable" upper bound for the number of taxis in the city.

What about a lower bound? Assuming the same probability model as before, we are certain that T is at least equal to 440, since we actually observed taxi #440. On the other hand, if $T = 440$, the probability that we would observe taxi #440 in a sample of 5 is only .011 (see Problem 4). So presumably T is larger than 440. Using the kind of argument that led to the upper bound 800 and computations that are only slightly more complicated, we find that a reasonable lower bound for T is 444, in symbols, $444 \leq T$ (see Problem 5). By combining our two statements, on the basis of what we have observed we can be reasonably confident that the true number of taxis is somewhere between 444 and 800,

$$444 \leq T \leq 800.$$

Such a statement is called a *confidence interval* for the quantity T. Confidence intervals are also referred to as *interval* estimates in contrast to the *point* estimates discussed in Chapter 1. When we estimated T as 529 in Chapter 1, we realized of course that this estimate deviated more or less from the true value T. There, we were in the fortunate position of being able to find out the true value of T. But in general that possibility does not exist. In such a case the point estimate does not provide any indication of the magnitude of its possible error. On the other hand a confidence interval has a built-in indicator of its accuracy, or possibly inaccuracy, namely the length of the interval. In our particular case the interval extends from 444 to 800, showing that a sample of 5 provides only very limited information about T.

PROBLEMS   1 The 1000 students in our imaginary college have indicated the following preferences for their area of concentration, where N denotes a concentration in the Natural Sciences, SS, in the Social Sciences, and H, in the Humanities.

area of concentration

		N	SS	H	totals
	F	75	125	100	300
	So	60	100	90	250
class	J	50	110	90	250
	S	45	85	70	200
	totals	230	420	350	1000

Find the following probabilities. (Always express verbally the event whose probability you are determining.)
a. $P(N)$, $P(SS)$, $P(H)$;
b. $P(F \text{ and } N)$, $P(J \text{ and } SS)$, $P(S \text{ and } H)$;
c. $P(L \text{ and } H)$, $P(U \text{ and } N)$;
d. $P(N \text{ or } SS)$, $P(L \text{ or } U)$;
e. $P(N \text{ or } S)$, $P(U \text{ or } H)$;
f. $P_S(N)$, $P_N(S)$, $P_S(H)$, $P_J(N \text{ or } SS)$, $P_U(H)$.

2 Consider the same information as for Problem 1.
a. Name several pairs of events that are exclusive.
b. Name two events that are independent.

3 Consider the first sample of taxi numbers in Table 1.1. Does 660 seem a reasonable value for the total number T of taxis? Why?

4 There are 440 taxis in a city, numbered from 1 to 440. A random sample of 5 taxis is observed. Show that the probability that taxi #440 appears at least once in the sample is .011. (*Hint:* Find the probability of the complementary event, that is, the event that #440 does not appear in the sample.)

5 There are 444 taxis in a city, numbered from 1 to 444.
   a. Show that the probability that in a random sample of 5 taxis the largest observed taxi number is 440 or greater is .055. (*Hint:* See Problem 4.)
   b. What is the probability of this event if there are only 443 taxis in the city?

6 Consider two-digit random numbers. Let E be the event that the number is divisible by 4; F, the event that the number is divisible by 5.
   a. Find $P(E$ and $F)$ and $P(E$ or $F)$.
   b. Are events E and F exclusive?
   c. Are events E and F independent?

7 If A and B are two exclusive events, with $P(A) = .3$ and $P(B) = .5$, find
   a. $P(A$ or $B)$;
   b. $P(A$ and $B)$;
   c. $P_A(B)$;
   d. $P_B(A)$.

8 Use a table of random digits such as Table J to find 5 taxi number samples of size 3 each, assuming that there are 75 taxis in all. For each of your samples compute the gap and median estimates. Which estimate would you say is closer to the true value?

9 What is the probability that among 5 random numbers, all known to lie between 1 and 51 (both limits included), none is smaller than 18?

10 How can a table of random numbers be used to simulate the pingpong ball experiment that produced the data for Table 3.6?

11 Two students are to be selected from the college population in Table 3.1. What is the probability that both are women if sampling is
   a. with replacement,
   b. without replacement?
   Compare the results with the corresponding results in Example 3.7. What can we conclude from this comparison?

4

Random
Variables*

Events and Simple Events

21 In Chapter 3 we used the term "event" in the usual colloquial
sense, and presumably the reader had no difficulty understanding
what was meant by any specific event. However, for mathematical
purposes such an intuitive approach is unsatisfactory. We must
say what we mean by event before we can talk rigorously of the
probability of an event.

We have been concerned with chance experiments, experiments
whose exact outcomes cannot be predicted with certainty. In
Chapter 3 the experiment usually consisted in the selection of one
student from among a given group of students. While we could not
predict which student would be selected, we could give a precise
listing of all possibilities. Quite generally then we assume that all
possibilities, or outcomes, associated with a chance experiment
are known, and we use the symbols e_1, e_2, . . . to denote the
possible outcomes of the experiment under discussion. Often the
number of possible outcomes is finite. In such a case we use the
symbol T to denote the total number of possible outcomes. On
the other hand there are many experiments for which it is necessary
or convenient to consider infinitely many outcomes. In such a
case it is, of course, physically impossible to list all outcomes.

We can now define what we mean by event. Any collection of
outcomes associated with a chance experiment is called an event.

---

*This chapter has been included for purposes of enrichment. In the remainder
of the book no specific reference is made to any of the material discussed in
this chapter (except in some of the starred problems). The mathematically
minded student who studies Chapter 4 should have little difficulty recognizing
its relevance to subsequent chapters.

In particular, a single outcome determines an event, often called a *simple* event. An arbitrary event can then also be considered as an aggregate of simple events.

A rigorous treatment of probability is relatively simple if there are only a finite number of outcomes. The case of infinitely many outcomes, particularly the noncountable case, raises additional mathematical problems, some of which are briefly mentioned in Chapter 12. To keep our treatment simple, we assume throughout the remainder of this chapter that the number of possible outcomes is finite. With the necessary mathematical preparation, relevant concepts can be redefined so that they apply also in the infinite case.

Consider then a random experiment with outcomes (or simple events) $e_1, \ldots, e_T$. Associated with each simple event e_j, $j = 1, \ldots, T$, is a nonnegative number p_j called the probability of the simple event e_j, $p_j = P(e_j)$, such that

$$p_1 + \cdots + p_T = 1.$$

If A is an arbitrary event, we set

$$P(A) = \sum_{e_j \text{ in } A} p_j,$$

where the summation extends over all simple events e_j that make up the event A.*

Random Variables and Probability Distributions

22   In practice we are often not interested in the particular outcome e_j of our experiment, but only in some special numerical feature associated with e_j. Thus in the taxi example of Chapter 1 the e_j are the actual taxis in town. But for our purposes we are only interested in one specific feature of the taxi, the number on its shield. Such features as color, make, or size are of no concern to us, at least for our particular problem.

An association that assigns a unique numerical value to every outcome of a chance experiment defines a *random* (or *chance*) *variable*. It is customary to use capitals X, Y, Z, etc. to denote random variables. However it is helpful to remember that when we write X we really mean $X = x(e)$, a (real-valued) function whose argument e is the outcome of the chance experiment under consideration.

Associated with a random variable X is a function $f(x)$ called the *probability distribution*, or simply, the distribution of X. For a given value x, $f(x)$ is the probability with which the random variable X takes the specific value x: $f(x) = P(X = x)$. Since we are considering only experiments with a finite number T of possible outcomes e_j, a random variable X can take at most T

---

*Students who are unfamiliar with summation notation should read Section 29.

different values. Let us denote these values by $x_1, \ldots, x_m$. Then for $i = 1, \ldots, m$, we have

4.1
$$f(x_i) = P(X = x_i) = \sum_{x(e_j)=x_i} p_j,$$

the sum of the probabilities of all simple events e_j for which the random variable X takes the value x_i. Since X takes one and only one value for each simple event e_j, we have

$$f(x_1) + \cdots + f(x_m) = p_1 + \cdots + p_T = 1.$$

EXAMPLE **Rolls of Two Fair Dice**   Our experiment consists in rolling
4.2a two fair dice. For definiteness suppose that one is red and the other green. The simple events of this experiment are the 36 number pairs $(1, 1)$, $(1, 2), \ldots, (r, g), \ldots, (6, 6)$, where the first number r in each pair indicates the number of points appearing on the red die, the second number g, the number of points appearing on the green die. The dice rolls are said to be *fair* if each of these 36 simple events has the same probability, namely $\frac{1}{36}$. For many dice games the only quantity of interest is the total number of points on both dice. If we denote this number by X, then X is a random variable defined for the simple events $e = (r, g)$, such that $X = x(e) = r + g$. The possible values of X are $2, 3, \ldots, 12$. In order to find the distribution of X we then need to know $f(x) = P(X = x)$ for $x = 2, 3, \ldots, 12$. For fair dice, simple enumeration gives the distribution

4.3

x	2	3	4	5	6	7	8	9	10	11	12
$f(x)$	$\frac{1}{36}$	$\frac{2}{36}$	$\frac{3}{36}$	$\frac{4}{36}$	$\frac{5}{36}$	$\frac{6}{36}$	$\frac{5}{36}$	$\frac{4}{36}$	$\frac{3}{36}$	$\frac{2}{36}$	$\frac{1}{36}$

EXAMPLE Consider trials with only two possible outcomes called success
4.4 and failure, respectively. Let K equal the number of successes in n independent trials. Then K is a random variable with possible values $k = 0, 1, \ldots, n$. In Chapter 5 we derive the probability distribution of the random variable K. (In Chapter 5 we write $b(k) = P(K = k)$, rather than $f(k)$.)

EXAMPLE **Random Variables and Statistical Inference**   We can
4.5 illustrate how the concept of random variable enters into statistical inference by looking once more at the taxi problem. Consider the experiment that consists in the random selection of one taxi from among the T taxis in town. The possible outcomes $e_1, \ldots, e_T$ of this experiment are the various taxis in town. We can then define a random variable W such that $W = w(e)$ equals the number on the taxi shield. The distribution of W is given by $f(w) = P(W = w) = 1/T$ for $w = 1, 2, \ldots, T$. In the taxi problem the number T is unknown. A constant like T associated with the distribution of a random variable is called a *parameter* of the distribution of the random variable. In Chapters 1 and 3 we

discussed how to estimate the parameter T of the taxi problem by means of a point estimate as well as by means of a confidence interval on the basis of n observed values for the random variable W.

The Expected Value of a Random Variable

23 A random variable X is characterized by its probability distribution. However, instead of knowing the complete distribution of X, it is often sufficient and even desirable to have only some summary information about the distribution. We usually want to know some typical or central value and perhaps to have some indication by how much individual values of X are likely to deviate from this typical value. There is no unique way to determine the center of a distribution. It is possible to generalize the concept of median, so that it applies to random variables and their distributions. (This is done in Section 77.) Now we define a quantity called the *mean* or *expected value*, denoted variously by the symbols μ, μ_x, or $E(X)$ (read, the expected value of X).

The mean or expected value of a random variable X is defined as the weighted sum

4.6
$$E(X) = \mu = x_1 f(x_1) + \cdots + x_m f(x_m) = \sum_i x_i f(x_i),$$

where again m is the number of different values the random variable X can take.

EXAMPLE
4.2b
For the sum of points of two fair dice we find from (4.3),
$$\mu = 2 \times \tfrac{1}{36} + 3 \times \tfrac{2}{36} + \cdots + 12 \times \tfrac{1}{36} = 7.$$

The significance of μ as a typical or central value becomes clear when we consider what happens when we repeat the chance experiment producing a value of X a large number of times and average the resulting values. Let $\#(x_i)$, $i = 1, \ldots, m$, stand for the number of times we observe the value x_i in n repetitions of the experiment. The average of all observed x-values (the term "average" understood in the sense of the arithmetic mean) is given by

$$\text{average} = \frac{x_1 \#(x_1) + \cdots + x_m \#(x_m)}{n}$$
$$= x_1 \frac{\#(x_1)}{n} + \cdots + x_m \frac{\#(x_m)}{n}$$
$$\doteq x_1 f(x_1) + \cdots + x_m f(x_m) = \mu,$$

according to the frequency interpretation of probability. Exactly as $f(x_i)$ may be looked upon as the long-run relative frequency of the value x_i, the mean μ may be looked upon as the long-run average value of X. If X is the amount of money a gambler wins or loses in a game of chance, μ is the average amount of money he can *expect* to win or lose per game in a large number of games.

The term *expected value* goes back to the time when the primary application of probability was to games of chance.

While the mean μ may be considered a typical value of X in the sense described above, it should be noted that μ need not be equal to any of the possible values of X (see Problem 3).

24   The concept of expected value can be extended. If $t(x)$ is a function of the argument x and X is a random variable, then $t(X)$ is a random variable, and its expected value is given by the weighted sum

$$E(t(X)) = \sum_i t(x_i)f(x_i)$$

where as before $f(x_i) = P(X = x_i)$.

A particular simple and useful result is obtained when $t(x)$ is a linear function; that is, $t(x) = a + bx$, where a and b are constants. Then

$$\begin{aligned} E(a + bX) &= \sum_i (a + bx_i)f(x_i) \\ &= \sum_i af(x_i) + \sum_i bx_i f(x_i) \\ &= a\sum_i f(x_i) + b\sum_i x_i f(x_i) = a + bE(X). \end{aligned}$$

We have proved the following important result:

THEOREM 4.7   *The expected value of a linear function of a random variable equals the linear function of the expected value of the random variable.*

The Variance

25   There is no unique way for measuring the degree of deviation of a random variable X from its mean μ. One very useful measure of deviation is the *variance* denoted by σ^2, σ_X^2, or $V(X)$. The variance of a random variable X is defined as the expected value of $(X - \mu)^2$,

$$\sigma^2 = E(X - \mu)^2 = \sum_i (x_i - \mu)^2 f(x_i).$$

EXAMPLE 4.2c   For the sum of points of two fair dice we find

$$\sigma^2 = (2 - 7)^2 \times \tfrac{1}{36} + (3 - 7)^2 \times \tfrac{2}{36} + \cdots + (12 - 7)^2 \times \tfrac{1}{36} = \tfrac{35}{6}.$$

By an analysis similar to the one involving the mean, we note that the variance may be considered as the average value of squared deviations from the mean μ in infinitely many experiments. It follows that the variance measures variability about the mean. A small value of σ^2 indicates that large deviations from the mean are very unlikely to occur, while a large value of σ^2 indicates that large deviations from the mean are not only possible but also likely to occur.

For practical purposes it is in general more useful to measure variability in terms of a quantity expressed in the same dimensionality as the actual observations (rather than the square of the observations). Thus for practical purposes the positive square root of the variance, the *standard deviation* denoted by σ, is often used as a measure of variability about the mean.

The definition of σ^2 given above is primarily useful for theoretical purposes. For purposes of computation another form is often preferable. With the help of Theorem 4.7 we find that

$$\begin{aligned} \sigma^2 = E(X - \mu)^2 &= E(X^2 - 2\mu X + \mu^2) \\ &= E(X^2) - 2\mu E(X) + \mu^2 \\ &= E(X^2) - 2\mu^2 + \mu^2 \\ &= E(X^2) - \mu^2. \end{aligned}$$

EXAMPLE 4.2d

Recomputation for the dice example gives

$$\sigma^2 = (2^2 \times \tfrac{1}{36} + 3^2 \times \tfrac{2}{36} + \cdots + 12^2 \times \tfrac{1}{36}) - 7^2 = \tfrac{35}{6}$$

as before.

Joint and Marginal Distributions of Two Random Variables

26   It often happens that two (or more) random variables are defined for each possible outcome of the same chance experiment. Thus for rolls of a red and a green die let R be the number of points showing on the red die, G the corresponding number for the green die. Then as on earlier occasions we may be interested in the random variable $X = R + G$. However we may also want to study the random variable $Y = |R - G| = |G - R|$, the numerical difference (without regard to sign) between the number of points on the individual dice. In particular, we may want to answer questions like, what is the probability that the sum of the points is 7 and the difference is 3? To answer this and similar questions, we introduce a new concept, the *joint* distribution of two random variables.

Let X be a random variable with possible values $x_i, i = 1, \ldots, m$, and let Y be a random variable with possible values $y_j, j = 1, \ldots, n$. The *joint probability distribution*, or simply the *joint distribution*, of X and Y is a function $h(x, y)$ such that

$$h(x_i, y_j) = P(X = x_i \text{ and } Y = y_j).$$

Such a function can be conveniently exhibited in a two-way table. (See page 32.)

The quantity $f(x_i) = \sum_j h(x_i, y_j)$ in the right-hand margin gives the probability $P(X = x_i)$ regardless of the value of Y. Similarly the quantity $g(y_j) = \sum_i h(x_i, y_j)$ in the lower margin gives the probability $P(Y = y_j)$ regardless of the value of X. The two functions $f(x)$ and $g(y)$ are usually called the *marginal* distributions of X and Y, respectively.

Joint and Marginal Distributions of X and Y

X \ Y	y_1	$\cdots$	y_j	$\cdots$	y_n	
x_1	$h(x_1, y_1)$	$\cdots$	$h(x_1, y_j)$	$\cdots$	$h(x_1, y_n)$	$f(x_1)$
$\vdots$	$\vdots$		$\vdots$		$\vdots$	$\vdots$
x_i	$h(x_i, y_1)$	$\cdots$	$h(x_i, y_j)$	$\cdots$	$h(x_i, y_n)$	$f(x_i)$
$\vdots$	$\vdots$		$\vdots$		$\vdots$	$\vdots$
x_m	$h(x_m, y_1)$	$\cdots$	$h(x_m, y_j)$	$\cdots$	$h(x_m, y_n)$	$f(x_m)$
	$g(y_1)$	$\cdots$	$g(y_j)$	$\cdots$	$g(y_n)$	

EXAMPLE 4.8 Consider the experiment of rolling a fair die. Two random variables X and Y are defined as follows:

X = remainder after dividing number of points on die by 2,
Y = remainder after dividing number of points on die by 3.

If e_j, $j = 1, \ldots, 6$, stands for the simple event that the die shows j points, we have the following setup.

	simple event					
	e_1	e_2	e_3	e_4	e_5	e_6
probability of simple event	$\frac{1}{6}$	$\frac{1}{6}$	$\frac{1}{6}$	$\frac{1}{6}$	$\frac{1}{6}$	$\frac{1}{6}$
$X = x(e)$	1	0	1	0	1	0
$Y = y(e)$	1	2	0	1	2	0

The joint distribution of X and Y is then given by:

X \ Y	0	1	2	
0	$\frac{1}{6}$	$\frac{1}{6}$	$\frac{1}{6}$	$\frac{1}{2}$
1	$\frac{1}{6}$	$\frac{1}{6}$	$\frac{1}{6}$	$\frac{1}{2}$
	$\frac{1}{3}$	$\frac{1}{3}$	$\frac{1}{3}$	

The concept of independence introduced in Chapter 3 is immediately applicable to random variables. Two random variables X and Y are said to be independent if for all possible combinations (x_i, y_j) we have

$$P(X = x_i \text{ and } Y = y_j) = P(X = x_i)P(Y = y_j),$$

or, in terms of distribution functions,

4.9
$$h(x_i, y_j) = f(x_i)g(y_j).$$

Thus, in the preceding example X and Y are independent random variables.

Expected Value and Variance of Sums of Two Random Variables

27 If we substitute (4.1) in (4.6), we find that the expected value of a random variable X can also be computed as

$$E(X) = x(e_1)p_1 + \cdots + x(e_T)p_T.$$

Similarly if X and Y are two random variables, we have

$$
\begin{aligned}
E(X + Y) &= [x(e_1) + y(e_1)]p_1 + \cdots + [x(e_T) + y(e_T)]p_T \\
&= x(e_1)p_1 + \cdots + x(e_T)p_T + y(e_1)p_1 + \cdots + y(e_T)p_T \\
&= E(X) + E(Y).
\end{aligned}
$$

We have proved the following important result:

THEOREM 4.10 *The expected value of the sum of two random variables equals the sum of the expected values of the two random variables.*

EXAMPLE 4.2e The random variable X of Example 4.2a can be written as $X = R + G$, where R equals the number of points on the red die and G equals the number of points on the green die. Then by Theorem 4.10,

$$E(X) = E(R) + E(G) = \tfrac{7}{2} + \tfrac{7}{2} = 7,$$

as in Example 4.2b.

28 We now turn to the expectation of the product of two random variables X and Y. Problem 14 shows that it is not generally true that the expected value of the product of two random variables equals the product of the expected values. However the statement is true for independent random variables:

THEOREM 4.11 *If X and Y are two independent random variables, then the expected value of the product XY equals the product of the expected values,*

$$E(XY) = E(X)E(Y).$$

Indeed, using (4.6) and (4.9),

$$
\begin{aligned}
E(XY) &= \sum_i \sum_j x_i y_j h(x_i, y_j) \\
&= \sum_i \sum_j x_i y_j f(x_i) g(y_j) \\
&= \left[\sum_i x_i f(x_i)\right]\left[\sum_j y_j g(y_j)\right] \\
&= E(X)E(Y).
\end{aligned}
$$

Theorem 4.11 allows us to prove the following important result:

THEOREM 4.12 *If X and Y are two independent random variables, then the variance of the sum $X + Y$ equals the sum of the individual variances,*

$$V(X + Y) = V(X) + V(Y).$$

By the definition of variance and Theorem 4.10, we have

$$\begin{aligned}
\text{var}(X + Y) &= E[(X + Y) - \mu_{x+y}]^2 \\
&= E[(X + Y) - (\mu_x + \mu_y)]^2 \\
&= E[(X - \mu_x) + (Y - \mu_y)]^2 \\
&= E(X - \mu_x)^2 + E(Y - \mu_y)^2 + 2E[(X - \mu_x)(Y - \mu_y)] \\
&= \text{var } X + \text{var } Y + 2C
\end{aligned}$$

where $C = E[(X - \mu_x)(Y - \mu_y)]$ is the so-called *covariance* of X and Y. We have to show that for independent random variables, $C = 0$. This result follows immediately when we apply Theorem 4.11 to the result of Problem 15.

EXAMPLE 4.2f   The computations required in Example 4.2c can be considerably simplified by writing $X = R + G$ and applying Theorem 4.12. Then $V(X) = V(R) + V(G) = \frac{35}{12} + \frac{35}{12} = \frac{35}{6}$ as before.

Theorems 4.10 and 4.12 immediately extend to sums of more than two random variables.

Summation Notation

29   In statistical work it is frequently necessary to compute sums of numbers. It is then often convenient to use "mathematical shorthand" to indicate the summation. Thus if $u_1, u_2, \ldots, u_m$ are m numbers, we shall write

$$\sum_i u_i$$

(read, "summation over i of u_i", or more briefly, "summation u_i") in place of the sum $u_1 + u_2 + \cdots + u_m$. The index i stands for the various subscripts from 1 to m to be included in the summation.

A few additional examples will illustrate the summation notation. Let $v_1, v_2, \ldots, v_m$ be a second set of m numbers and let a and b be constants. We shall consider expressions such as

$$\sum_i u_i v_i = u_1 v_1 + u_2 v_2 + \cdots + u_m v_m$$

and

$$\begin{aligned}
\sum_i (au_i + bv_i) &= (au_1 + bv_1) + (au_2 + bv_2) + \cdots + (au_m + bv_m) \\
&= a(u_1 + u_2 + \cdots + u_m) + b(v_1 + v_2 + \cdots + v_m) \\
&= a \sum_i u_i + b \sum_i v_i.
\end{aligned}$$

Or again, let $x_1, x_2, \ldots, x_m$ and $y_1, y_2, \ldots, y_n$ be two sets of numbers, where m may or may not equal n. We write

$$\sum_i \sum_j x_i y_j$$

to indicate summation over all indices i from 1 to m and all indices j from 1 to n. Thus

$$\sum_i \sum_j x_i y_j = x_1(y_1 + y_2 + \cdots + y_n) + x_2(y_1 + y_2 + \cdots + y_n)$$
$$+ \cdots + x_m(y_1 + y_2 + \cdots + y_n)$$
$$= (x_1 + x_2 + \cdots + x_m)(y_1 + y_2 + \cdots + y_n)$$
$$= \left(\sum_i x_i\right)\left(\sum_j y_j\right).$$

When we write $\sum_i$ or $\sum_j$ it is understood that summation extends over all possible subscripts i or j, respectively. Sometimes, as in the definition of $P(A)$ in Section 21, we want to sum over only a selected set of subscripts. Thus when computing $P(A)$ we sum only over subscripts j such that the event A is satisfied for the outcome e_j. We indicate this by writing

$$\sum_{e_j \text{ in } A} p(e_j).$$

PROBLEMS

1 Consider the random variable K of Example 4.4. What are the experimental outcomes e that determine the value of K? How many such outcomes are there?

For Problems 2 through 8 use the following information. Let R equal the number of points on a red die, G, those on a green die. Always assume that dice tosses are fair.

2 Write out the distribution of R.

3 Show that $E(R) = \frac{7}{2}$.

4 Show that $E(R - \frac{7}{2}) = 0$.

5 Show that var $R = \frac{35}{12}$.

6 If in Problems 2 through 5, R is replaced by G, what are the corresponding answers? Why?

7 Find the joint distribution of $X = R + G$ and $Y = |G - R|$.

8 Find the marginal distribution of X in Problem 7. Compare your result with the distribution of X in Example 4.2a.

9 If X is any random variable, show that $E(X - \mu) = 0$. It follows that $E(X - \mu)$ is useless as a measure of variability of X about its mean.

10 Let c be a constant. Show that $E(X - c)^2$ is a minimum when $c = E(X) = \mu$. (*Hint:* Write $E(X - c)^2 = E[(X - \mu) + (\mu - c)]^2$.)

11 If the random variable X has variance σ_x^2 and if a and b are constants, show that the random variable $Y = a + bX$ has variance $\sigma_y^2 = b^2\sigma_x^2$.

12 Let I be a random variable that takes only the values 1 and 0 with probabilities p and $q = 1 - p$, $0 < p < 1$. (Such a variable is often called an *indicator variable*.) Show that
a. $E(I) = E(I^2) = p$,
b. $V(I) = pq$.

13 Let $I_1, \ldots, I_n$ be a set of independent random variables, each having the same distribution as the random variable I of Problem 12. Let $K = I_1 + \cdots + I_n$.

a. Give a verbal interpretation of the meaning of the random variable K.
b. Find $E(K)$.
c. Find $V(K)$.

14 The random variable U takes the values $+1$ and -1 with probability $\frac{1}{2}$ each.
a. Find $E(U)$.
b. Find $E(U^2)$.
c. Use the results of (a) and (b) to show that

$$E(U^2) = E(U \times U) \neq E(U) \times E(U),$$

so that in this case the expected value of a product does not equal the product of the expected values.

15 Let X and Y be two random variables. Show that we have the following computing formula for the covariance of X and Y:

$$E[(X - \mu_x)(Y - \mu_y)] = E(XY) - \mu_x\mu_y.$$

5

The Binomial Distribution

30   In this and some of the following chapters, we consider experiments consisting of a sequence of trials in each of which a certain result does or does not occur. A card player does or does not get a hand containing 2 aces. A single random digit is or is not greater than 4. The Boston Red Sox do or do not win a given game with the Yankees. Clearly there is no end to examples of this type. We start with probabilistic problems arising in connection with sequences of trials. After that we take up statistical problems. But first of all we must introduce some basic terminology and notation.

Almost universally in the mathematical and statistical literature, the two possible outcomes of a trial are given the names *success* and *failure*. In everyday life these two words have very specific meanings. In probability and statistics they are simply convenient names without the usual connotations. It is quite possible that the death of a patient is called a "success" if the investigator should be studying the fatality rate associated with a certain kind of operation. We denote the number of trials to be performed in a given experiment by n, and it is always understood that the value of n has been fixed before the start of the whole experiment. Thus the theory that we are going to develop does not apply if n is determined in the course of the experiment, as in the case of the gambler who looks at his past winnings or losses and decides that he has had enough. We also always assume that successive trials are independent of each other, and that the probability of success remains the same throughout the experiment. There are many experiments where one or both of these assumptions are violated. In such cases the theory that we are going to develop, if applied, may give quite erroneous answers.

We always use the letter p for the probability of success in a single trial. For the probability of failure we write q. Since a trial

37

has only two possible outcomes, success and failure, we must have $q = 1 - p$. Thus we have the following notation:

n = number of trials

p = probability of success in any one trial

$q = 1 - p$ = probability of failure in any one trial.

Binomial Probabilities

31 Now we can restate the problem we want to solve. If k is any integer between 0 and n, what is the probability $b(k)$ that in an experiment consisting of n trials, k of the trials are successful and $n - k$ are unsuccessful? However before we tackle this general problem, let us first look at a special case. At a country fair we win a prize if we roll either a five or a six with a single die. If we decide to try our luck three times, what is the probability that we will win exactly two prizes? If the die we are using is fair, we have two chances in six of winning a prize every time we roll the die. Let us agree to call winning a prize a success and not winning a prize a failure. We then want to solve the following problem: What is the probability of observing two successes and one failure in three trials with success probability $\frac{1}{3}$ at each trial?

It is a good idea to start by making a list of all the possible ways in which we can have two successes. The best way of doing this is to write down what happens on each of the 3 trials, letting S represent success and F, failure. This list is quite simple:

$$SSF \quad SFS \quad FSS$$

Our next problem is to find the probability of each of these three possibilities. Once we have done this, the final step is simple. Since only one of the three possibilities can happen at any one time, we simply add the three probabilities.

We take the case SSF first. Since we assume that successive trials are independent (what happens at one roll of the die has no influence on what happens at the other rolls of the die), we have

$$P(SSF) = P(S)P(S)P(F) = \tfrac{1}{3} \times \tfrac{1}{3} \times \tfrac{2}{3} = \tfrac{2}{27}.$$

Similarly

$$P(SFS) = P(S)P(F)P(S) = \tfrac{1}{3} \times \tfrac{2}{3} \times \tfrac{1}{3} = \tfrac{2}{27},$$

and

$$P(FSS) = P(F)P(S)P(S) = \tfrac{2}{3} \times \tfrac{1}{3} \times \tfrac{1}{3} = \tfrac{2}{27}.$$

The sum of these three probabilities is $\frac{6}{27} = \frac{2}{9}$, and this is the probability of winning exactly two prizes in our three attempts.

32 In exactly the same fashion we could compute, say, the probability of winning one prize in three trials. But let us do something more general. We have computed our probability under the specific assumption that $p = \frac{1}{3}$, that is, that we are using a fair die. But suppose we want to know our chances if p is equal to .30 or .28 rather than .33. (Not all dice are necessarily fair.) Clearly our method works for any specific value of p, not only for $p = \frac{1}{3}$. Indeed there is no reason at all why we cannot use the letter p instead of a numerical probability. We can actually use part of our earlier results:

$$P(SSF) = p \times p \times q = p^2q$$
$$P(SFS) = p \times q \times p = p^2q$$
$$P(FSS) = q \times p \times p = p^2q.$$

All three arrangements have the same probability p^2q. The reason for this is quite obvious. Each arrangement contains the letter S twice and the letter F once. And since according to our assumption successive trials are independent of each other, each S contributes p and each F contributes q to the probability of the given arrangement. Thus the probability we are looking for is equal to the probability of any one of the possible arrangements, multiplied by the number of possible arrangements, in our case, 3:

$$P(2 \text{ prizes}) = 3p^2q.$$

If we set $p = \frac{1}{3}$ and $q = \frac{2}{3}$, we should get our earlier result. Indeed, $3(\frac{1}{3})^2(\frac{2}{3}) = \frac{2}{9}$.

Let us now do the same for 1 win and 2 losses. A typical arrangement giving this result is SFF and

$$P(SFF) = P(S)P(F)P(F) = pqq = pq^2.$$

There are two other arrangements, namely FSF and FFS, that have the same probability. Thus

$$P(1 \text{ prize}) = 3pq^2.$$

To complete the picture,

$$P(3 \text{ prizes}) = P(SSS) = P(S)P(S)P(S) = ppp = p^3,$$
$$P(\text{no prize}) = P(FFF) = q^3.$$

Here we have brought together our results:

number of successes k	probability $b(k)$ of k successes	special case: $p = \frac{1}{3}$
0	q^3	$\frac{8}{27}$
1	$3pq^2$	$\frac{4}{9}$
2	$3p^2q$	$\frac{2}{9}$
3	p^3	$\frac{1}{27}$

The probabilities in the middle column are the terms in the expansion of $(q + p)^3$. Since an expression like $q + p$ is often called a binomial, the corresponding probabilities are called *binomial probabilities*. Together they form the *binomial distribution*. In particular, in our case we speak of the binomial distribution with $n = 3$ and success probability p.

Let us use our results to solve the following problem. What is the probability that we win at least one prize? Since at least one means 1, 2, or 3 prizes, our probability is $b(1) + b(2) + b(3)$. We can simply substitute in this expression from our earlier table of probabilities. Actually, in this case it is possible to give a more compact answer. The sum of all 4 probabilities must be one, since 0, 1, 2, and 3 are the only possibilities for the number of successes in 3 trials. Therefore

$$P(\text{at least one prize}) = 1 - b(0) = 1 - q^3.$$

In particular, for an honest die we have

$$1 - (\tfrac{2}{3})^3 = 1 - \tfrac{8}{27} = \tfrac{19}{27} \text{ or } .70.$$

A General Formula

33 We are now ready to solve the general problem. What is the probability $b(k)$ that in n trials with probability p of success in a single trial, exactly k of the trials end in success and $n - k$ end in failure? As we have seen we need only multiply the probability of an arbitrary arrangement of k successes and $n - k$ failures by the number of possible arrangements. One such arrangement is $SS \ldots SFF \ldots F$, a sequence of k successes followed by a sequence of $n - k$ failures, having probability

$$P(SS \ldots SFF \ldots F) = pp \cdots pqq \cdots q = p^k q^{n-k}.$$

It follows that

$$b(k) = B(n, k)p^k q^{n-k},$$

where the so-called *binomial coefficient* $B(n, k)$ tells us in how many different ways k letters S and $n - k$ letters F can be arranged.

The derivation of an appropriate formula for $B(n, k)$ is not difficult but it is quite time-consuming. Since many readers probably have seen a derivation before or will see one in another course, we shall simply write down the appropriate formula, namely,

$$B(n, k) = \frac{n!}{k!(n - k)!}$$

where $n! = n(n - 1)(n - 2) \cdots (3)(2)(1)$ and $0! = 1$. For example, $5! = (5)(4)(3)(2)(1) = 120$. We then have

5.1
$$b(k) = \frac{n!}{k!(n - k)!} p^k q^{n-k}.$$

As a check, let us take $n = 3$, $k = 2$, and $p = \tfrac{1}{3}$. Then

$$b(2) = \frac{3!}{2!1!} (\tfrac{1}{3})^2 (\tfrac{2}{3}) = \tfrac{2}{9},$$

which agrees with the result we obtained in our special problem involving the probability of winning two prizes in three trials with a fair die.

Theoretically we can use our formula to compute the probability of k successes for any values of k, n, and p. But there are practical difficulties. The reader should try to compute a few of these probabilities for various values of k, n, and p in order to appreciate the amount of work that is required! Of course with high-speed computers this is no problem. Statisticians use ready-made tables whenever possible. Since a statistician can never tell which values of n and p he is going to encounter, tables of the binomial distribution have to be very extensive.

Table A contains binomial probabilities for certain selected values of n and p. These tables are for illustrative purposes only. For extensive statistical work the student will have to consult special tables of the binomial distribution or, where appropriate, use the normal approximation to be discussed in the next chapter.

EXAMPLE
5.2
What is the probability of observing 4 successes in 10 trials with success probability .20? According to Formula 5.1,

$$b(4) = \frac{10!}{4!6!} \cdot .2^4 \times .8^6 = .0881.$$

More conveniently, to three-place accuracy, the answer can be found in Table A as follows:

(i) go to the table for $n = 10$,

(ii) find the column labeled $p = .2$,

(iii) if—as in this case—p is listed in the *top* row, look for the appropriate k-value in the *left-hand* margin. The desired probability is listed at the intersection of the row labeled $k = 4$ and the column labeled $p = .2$: $b(4) = .088$.

EXAMPLE
5.3
What is the probability of observing 16 successes in 20 trials with success probability .70? We use Table A for $n = 20$. The value $p = .70$ is found in the *bottom* row. Therefore we look for k in the *right-hand* margin. At the intersection of the row labeled $k = 16$ and the column labeled $p = .70$ we find $b(16) = .130$.

EXAMPLE
5.4
What is the probability of observing at least 16 successes in 20 trials with success probability .70? We have
P (at least 16 successes)

$$
\begin{aligned}
&= P(k \geq 16) \\
&= b(16) + b(17) + b(18) + b(19) + b(20) \\
&= .130 + .072 + .028 + .007 + .001 \\
&= .238.
\end{aligned}
$$

PROBLEMS

1 Find $b(k)$ for $n = 3$, $p = \frac{1}{6}$, $k = 0, 1, 2, 3$.

2 Find $b(k)$ for $n = 10$, $p = \frac{1}{3}$, $k = 0, 1, \ldots, 10$.

3 If you roll a fair die three times, what is the probability that you observe at least one "6"?

4 In 10 binomial trials with success probability $\frac{1}{3}$, find the probability of
a. at least 5 successes,
b. at most 2 successes,
c. more than 4 successes,
d. fewer than 5 successes.

5 In 10 binomial trials with success probability $\frac{2}{3}$, find the probability of
a. exactly 6 successes,
b. more than 6 successes,
c. fewer than 6 successes,
d. at least 8 successes.

6 An examination consists of 25 questions, at least 17 of which have to be answered correctly for a passing grade. A student knows the answers to 60% of the type of questions likely to be on the examination. What is his probability of receiving a passing grade?

7 A certain do-it-yourself job requires the use of 12 screws. It has been found that approximately 10% of all kits required for the job contain at least one faulty screw. A store sells 25 of the kits. Find the probability that the store will receive
a. no complaint about faulty screws,
b. at most 2 complaints.
What assumption(s) is (are) implied in your answer?

6

The Normal Distribution

An Approximation for Binomial Probabilities

34   It is often said that one picture is worth a thousand words. Most mathematicians feel that a mathematical formula is even better than a picture. However, there are times when pictures and graphs are very useful in interpreting concretely what a formula has to say. For example, consider the formula for binomial probabilities that we discussed in Chapter 5,

$$b(k) = \frac{n!}{k!(n-k)!} p^k (1-p)^{n-k}, k = 0, 1, \ldots, n.$$

In particular for $n = 3$ and $p = \frac{1}{3}$ we can compute the successive probabilities

$$b(0) = \tfrac{8}{27},$$
$$b(1) = \tfrac{12}{27},$$
$$b(2) = \tfrac{6}{27},$$
$$b(3) = \tfrac{1}{27}.$$

But a picture presents much more vividly how these probabilities change with k. The customary way to present such probabilities is by means of a *histogram*. We plot the possible values of k along a horizontal axis and then erect rectangles with base centered at k and height equal to $b(k)$ as in Figure 6.1. Since each rectangle has a base of length 1, the areas of the various rectangles are simply our probabilities $b(k)$. In a histogram the eye automatically concentrates on areas and we get a very graphic picture of the probability distribution, that is, the way in which the total probability 1 is distributed among the possible values of k.

FIGURE
6.1

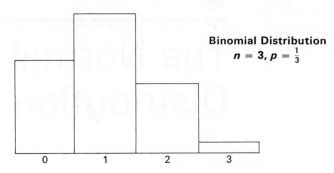

Binomial Distribution
$$n = 3, p = \tfrac{1}{3}$$

It is interesting to study how a change in the number of trials affects the probabilities $b(k)$. In many courses on probability a great deal of time and effort is expended on such studies, resulting in important theorems. In this course we are content to use the results of such efforts. Before we do so, we want to indicate graphically how these results come about. For this purpose we look at two more histograms of binomial distributions, again with success probability $\tfrac{1}{3}$ but with different values of n. In Figure 6.2

FIGURE
6.2

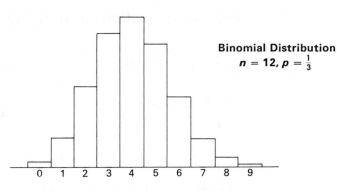

Binomial Distribution
$$n = 12, p = \tfrac{1}{3}$$

FIGURE
6.3

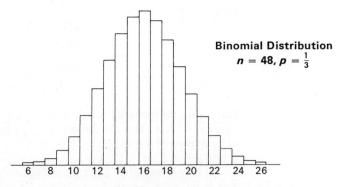

Binomial Distribution
$$n = 48, p = \tfrac{1}{3}$$

we have $n = 12$ and in Figure 6.3, $n = 48$. As we go from 3 to 12 to 48 trials, we note a striking fact. The histograms smooth out and begin to take on a more symmetric shape. In courses dealing with probability it is shown by advanced calculus methods that this process continues as the number of trials increases, and that eventually the histograms tend to the so-called *normal curve* pictured in Figure 6.4. If we cut out the normal curve and super-impose it on the histograms (centering it at the points 1, 4, and 16, respectively), we find that it gives a rather bad fit for $n = 4$, a considerably improved fit for $n = 12$, and a very satisfactory fit for $n = 48$.

FIGURE 6.4

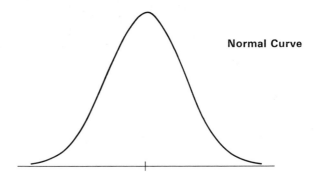

Normal Curve

Our reasons for studying these results are primarily practical. Suppose we are interested in computing the probability that in 48 trials with success probability $\frac{1}{3}$ we observe between 10 and 20 successes, both limits included. From an appropriate table we find

$$P(10 \leq k \leq 20) = b(10) + \cdots + b(20) = .894.$$

But suppose an appropriate table is not available. We might try to compute each one of the 11 probabilities from our formula. But this is a rather hopeless task, unless we have a high-speed computer at our disposal. What else can we do? It turns out that our new discovery provides us with a very simple means of com-puting the desired probability, at least approximately. A look at Figures 6.3 and 6.4 shows how. We know that each probability $b(k)$ is represented by the area of a rectangle. This means that the probability we are looking for is represented by the area of a sum of rectangles, namely those labelled 10, 11, and so on up to 20. Suppose now we try to approximate this area by the corresponding area under the normal curve. Since the normal curve seems to fit rather closely over the histogram, we should get a reasonably good approximation to the desired probability. Of course we would use the area under the curve between the lower endpoint of the rectangle on the left and the upper endpoint of the rectangle on the right, that is, the points 9.5 and 20.5. For future reference we note that $9.5 = 10 - \frac{1}{2}$ and $20.5 = 20 + \frac{1}{2}$, where 10 and 20 are the most extreme numbers of successes in the interval in which we are interested.

35   But how are we going to compute the area under the normal curve? In a course in calculus it is shown that areas under a curve are computed by means of integration. Fortunately we do not have to start integrating every time we want to use the normal curve, which we shall frequently want to do. Instead, we can use an appropriate table. Before we can describe how to use such a table, we have to make one additional remark. From our discussion so far it may appear that there is just one normal curve. This is not so. A normal curve depends on two constants which are called the *mean* and the *standard deviation*, customarily represented by the symbols μ (Greek mu) and σ (Greek sigma), respectively. The mean μ can be any number, the standard deviation σ, any positive number. The meaning of these two constants is best seen from Figure 6.5.

FIGURE 6.5

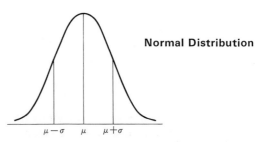

Normal Distribution

The mean μ simply tells us where the center of symmetry of the normal curve is. The meaning of the standard deviation σ is more complicated. Graphically σ is the distance from the center of symmetry to the two points at which the curve changes from curving down to curving up. A change in the value of μ does not change the outline of the curve. It simply moves the whole curve further to the right or left depending on whether μ increases or decreases. A change in σ, on the other hand, changes the outline of the curve. This curve is always symmetric about μ, but for small values of σ it has a high and narrow outline, and for large values of σ its outline is low and flat.

As we have mentioned, we are going to use areas under a normal curve to compute probabilities. Since the curve tells us how probability is distributed over various intervals, we shall usually talk of normal distributions rather than normal curves. More specifically, we shall talk of the normal distribution with mean μ and standard deviation σ. As we shall see, one normal distribution plays a special role in the computation of probabilities, the so-called *standard* or *unit* normal distribution. This is the normal distribution with mean zero and standard deviation one. In general we shall reserve the term normal curve for the standard normal distribution.

We have introduced the normal distribution as a means for approximating probabilities from binomial experiments. Actually the significance of the normal distribution goes much beyond this particular application. As we proceed in our study of statistics, we shall encounter the normal distribution over and over again. For this reason we study now in some detail how areas relating to a normal distribution are found.

Areas under the Normal Curve

36 Suppose we have a normal distribution with mean μ and standard deviation σ and want to find the area under this normal distribution between two numbers x_1 and x_2. It is shown by calculus methods that this problem is equivalent to the following problem. We compute the two numbers

$$z_1 = \frac{x_1 - \mu}{\sigma} \text{ and } z_2 = \frac{x_2 - \mu}{\sigma}$$

and then find the area under the normal curve, that is, the normal distribution with mean 0 and standard deviation 1, between z_1 and z_2. This area is the same as the area under the original normal distribution between x_1 and x_2. As a consequence, it is sufficient if we learn how to look up areas relating to the normal curve.

FIGURE 6.6

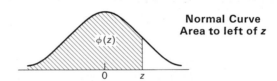

Normal Curve
Area to left of z

For any number z let us denote the area under the normal curve to the left of z by $\phi(z)$ as in Figure 6.6. The area under the normal curve between two numbers z_1 and z_2 (with z_1 the smaller number) is then given by

$$\phi(z_2) - \phi(z_1),$$

since the desired area is the difference between the area to the left of z_2 and the area to the left of z_1. (See Figure 6.7.)

FIGURE 6.7

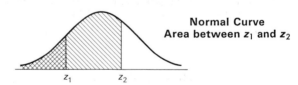

Normal Curve
Area between z_1 and z_2

Table B gives values of $\phi(z)$ (rounded to four decimal places) for positive z starting with $z = 0$ and increasing in steps of .01.

EXAMPLE 6.8

$\phi(1.23) = .8907$, the number given in Table B at the intersection of the row labeled 1.2 and the column labeled .03. (Note that 1.23 equals 1.2 + .03.)

EXAMPLE 6.9

The area under the normal curve between .87 and 1.87 equals $\phi(1.87) - \phi(.87) = .9693 - .8078 = .1615$.

In addition to the numerical information provided by Table B we need to know the following two properties of the normal curve in order to determine areas between two arbitrary numbers z_1 and z_2:

(i) The normal curve is symmetric with respect to $z = 0$.
(ii) The total area under the normal curve is one.

Properties (i) and (ii) imply two additional properties that are often helpful in finding areas under the normal curve.

(iii) The area under the normal curve to the right of z is $1 - \phi(z)$.
(iv) $\phi(-z) = 1 - \phi(z)$. That is, the area to the left of $-z$ equals the area to the right of z.

EXAMPLE 6.10

The area under the normal curve between $-.87$ and 1.87 is

$$\phi(1.87) - \phi(-.87) = \phi(1.87) - [1 - \phi(.87)]$$
$$= .9693 - .1922 = .7771.$$

EXAMPLE 6.11

The area under the normal curve between -1.87 and $-.87$ is

$$\phi(-.87) - \phi(-1.87) = [1 - \phi(.87)] - [1 - \phi(1.87)]$$
$$= \phi(1.87) - \phi(.87) = .1615$$

by Example 6.9. The same result is obtained more directly by realizing that in view of the symmetry of the normal curve the area between -1.87 and $-.87$ equals the area between $.87$ and 1.87.

EXAMPLE 6.12

The area under the normal curve to the right of $-.87$ is $1 - \phi(-.87) = 1 - [1 - \phi(.87)] = \phi(.87) = .8078$. Again a faster way is to realize that due to symmetry the area to the right of $-.87$ equals the area to the left of $.87$.

EXAMPLE 6.13

Interpolation   The area under the normal curve to the left of 1.024 equals $\phi(1.024)$. Table B only gives the values

$$\phi(1.02) = .8461,$$
$$\phi(1.03) = .8485.$$

Thus $\phi(1.024)$ is between $.8461$ and $.8485$. For most practical purposes it is sufficient to determine probabilities, that is, areas, to an accuracy of two or at most three decimal places. Mere inspection shows that we should take $\phi(.1024) = .847$. If needed, a more accurate value is obtained by linear interpolation as follows. Since 1.024 is $\frac{4}{10}$ of the way between 1.02 and 1.03, we approximate $\phi(1.024)$ as $\frac{4}{10}$ of the way between $\phi(1.02)$ and $\phi(1.03)$ or

$$\phi(1.024) = \phi(1.02) + \tfrac{4}{10}[\phi(1.03) - \phi(1.02)]$$
$$= .8461 + \tfrac{4}{10}(.8485 - .8461)$$
$$= .8461 + .0010$$
$$= .8471.$$

37   As we shall see, in statistical applications we often have the reverse problem. Rather than find the value of $\phi(z)$ for given z, we need to find the value z that produces a given value $\phi(z)$.

EXAMPLE
6.14

Find z such that $\phi(z) = .975$. In Table B at the intersection of the row labeled 1.9 and the column labeled .06 we find the number .9750. It follows that $\phi(1.96) = .9750$. Therefore $z = 1.96$.

EXAMPLE
6.15

Find z such that $\phi(z) = .3085$. Since $\phi(z) < \frac{1}{2}$, according to Properties (i) and (ii), z must be negative. We therefore consider $\phi(-z) = 1 - \phi(z) = .6915$. (Note that if z is negative, $-z$ is positive.) By the method of Example 6.14 we find $-z = .50$ and therefore, $z = -.50$.

EXAMPLE
6.16

Find z such that $\phi(z) = .75$. According to Table B, $\phi(.67) = .7486$ and $\phi(.68) = .7517$. It follows that z lies between .67 and .68. Since .75 is just about half-way between .7486 and .7517, for most practical purposes $z = .675$ is sufficiently accurate. (If two-place accuracy is sufficient, we use $z = .67$ since $\phi(.67)$ is closer to .75 than $\phi(.68)$.) Linear interpolation would proceed as follows:

$$z = .67 + \frac{.7500 - .7486}{.7517 - .7486} \times \frac{1}{100}$$

$$= .67 + .0045$$

$$= .6745.$$

EXAMPLE
6.17

Find z such that $\phi(z) = .025$. We have $.975 = 1 - \phi(z) = \phi(-z)$ by Property (iv). Therefore by the result of Example 6.14, $-z = 1.96$ or $z = -1.96$.

The next example illustrates another type of problem which arises frequently in statistical applications.

EXAMPLE
6.18

Find an interval that contains the central 95% of the area under the normal curve. In other words, we want to find the value z such that $\phi(z) - \phi(-z) = .95$. Formally we can proceed as follows. We replace $\phi(-z)$ by $1 - \phi(z)$ and obtain $2\phi(z) - 1 = .95$ or $\phi(z) = .975$. It then follows that $z = 1.96$ by Example 6.17. Actually in this, as well as in most earlier examples, a less formal approach is preferable. Since the area between $-z$ and z is to be .95, the area below $-z$ must be $\frac{1}{2}(.05) = .025$, so that the area below z equals $.025 + .95 = .975$ and $z = 1.96$ according to Table B. (See Figure 6.19.)

FIGURE
6.19

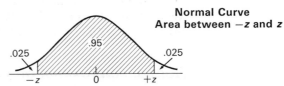

Normal Curve
Area between $-z$ and z

The problem illustrated in Example 6.18 arises so frequently in statistical applications that it is convenient to have a ready-made table of z-values for various central areas. Table C gives values z such that $\phi(z) - \phi(-z) = \gamma$ for certain selected values of γ occurring frequently in statistical applications. In addition to γ,

Table C also lists $1 - \gamma$ and $\frac{1}{2}(1 - \gamma)$. Thus Example 6.18 shows that the z-values in Table C have also the following property:

$$\text{(area above } z) = \text{(area below } -z) = \tfrac{1}{2}(1 - \gamma).$$

In particular, Example 6.16 could have been solved by reference to Table C with $\frac{1}{2}(1 - \gamma) = .25$; to three-place accuracy the answer is $z = .674$.

38 For Examples 6.20 and 6.21 we consider a normal distribution with mean 100 and standard deviation 10.

EXAMPLE 6.20

Find the area A between $x_1 = 80$ and $x_2 = 110$. We first compute

$$z_1 = \frac{x_1 - \mu}{\sigma} = \frac{80 - 100}{10} = -2$$

and

$$z_2 = \frac{x_2 - \mu}{\sigma} = \frac{110 - 100}{10} = 1.$$

It follows that $A = \phi(1) - \phi(-2) = .8413 - .0228 = .8185$.

EXAMPLE 6.21

Find the two x-values that contain the central 80% of the total area. From Table C we find that the central 80% under the normal curve falls between -1.282 and 1.282. Thus we have

$$-1.282 = z_1 = \frac{x_1 - 100}{10}$$

or

$$x_1 = 100 - 12.82 = 87.18$$

and

$$1.282 = z_2 = \frac{x_2 - 100}{10}$$

or

$$x_2 = 100 + 12.82 = 112.82.$$

Example 6.21 can be generalized as follows. The central $100\gamma\%$ of the area under a normal distribution with mean μ and standard deviation σ lies between $\mu - z\sigma$ and $\mu + z\sigma$, where z is read from Table C. The standard deviation of a normal distribution is the most useful unit for measuring deviations from the mean. When we compute $z = (x - \mu)/\sigma$, we find out how many standard deviations the point x is away from the mean μ.

Normal Approximation for Binomial Probabilities

39 Now we finally return to the problem that led to our discussion of the normal distribution in the first place. How can we approximate the probability of observing between 10 and 20 successes in 48 binomial trials with success probability $\frac{1}{3}$? Our earlier discussion showed that this probability can be approximated by the area under a suitable normal distribution between the points $x_1 = 9.5$ and $x_2 = 20.5$. It can be shown (see Problem 13) that the appropriate mean and standard deviation are np and $\sqrt{npq}$ respectively. In particular for $n = 48$ and $p = \frac{1}{3}$ we find $\mu = 48 \times \frac{1}{3} = 16$ and $\sigma = \sqrt{48 \times \frac{1}{3} \times \frac{2}{3}} = 3.266$. Thus

$$z_1 = \frac{9.5 - 16}{3.266} = -1.990, \quad z_2 = \frac{20.5 - 16}{3.266} = 1.378.$$

Finally from Table B we find that the area between -1.990 and 1.378 is $.893$. This is our approximation to the desired probability. It differs from the exact probability $.894$ obtained directly from the binomial distribution only slightly in the third decimal place.

Our procedure can of course be generalized. Let k_1 and k_2 be two integers satisfying the condition $0 \leq k_1 \leq k_2 \leq n$. We should like to approximate the probability that in n binomial trials with success probability p we observe between k_1 and k_2 successes, both limits included. Our previous discussion suggests that if the number of trials is sufficiently large, the area under the normal distribution with mean $\mu = np$ and standard deviation $\sigma = \sqrt{npq}$ between the points $x_1 = k_1 - \frac{1}{2}$ and $x_2 = k_2 + \frac{1}{2}$ provides a suitable approximation. Thus we have to compute the two values

$$z_1 = \frac{k_1 - \frac{1}{2} - np}{\sqrt{npq}} \quad \text{and} \quad z_2 = \frac{k_2 + \frac{1}{2} - np}{\sqrt{npq}}$$

and find the area between z_1 and z_2 from tables of the normal distribution. For most practical purposes the approximation is satisfactory if npq is at least 3. In particular this means that if the success probability p is close to $\frac{1}{2}$ the normal approximation, as it is called, can be used for values of n as small as 12. However if p is either very close to 0 or very close to 1, n has to be quite large. As an example, for $p = \frac{1}{100}$, n should be over 300 before we can expect the normal approximation to give reliable results.

If either $k_1 = 0$ or $k_2 = n$, the following simplification is recommended. When $k_1 = 0$, we set $\phi(z_1) = 0$. When $k_2 = n$, we set $\phi(z_2) = 1$.

EXAMPLE 6.22   Use the normal approximation to compute the probability that in 150 trials with success probability .6 we observe at least 100 successes. In this problem we have $k_1 = 100$ and $k_2 = n$. Since $\mu = 150 \times .6 = 90$ and $\sigma = \sqrt{150 \times .6 \times .4} = \sqrt{36} = 6$, the desired probability is approximated as

$$1 - \phi\left(\frac{100 - \frac{1}{2} - 90}{6}\right) = 1 - \phi(1.58) = .057$$

40   Let us look at one further example of the use of the normal approximation. Suppose we toss a fair penny 100 times, or what amounts to the same, drop 100 fair pennies on a table. In an experiment of this type we might expect to get 50 heads and 50 tails. But computations show that the probability of an even split is only .08 (see Problem 11), so that 92% of the time we get either more than or fewer than 50 heads. We may then ask between what limits the number of heads is likely to fluctuate. In this form the question is too vague for an answer. Strictly speaking the occurrence of 0 heads or 100 heads is not impossible, only extremely unlikely. We should ask instead, how large a deviation from an even split can reasonably be expected. Of course the word reasonable is also open to all sorts of interpretations. But statisticians are often willing to consider as reasonable deviations that occur about 95% of the time. Before you read on, try to guess what kind of deviations are reasonable when you toss a coin 100 times.

From Table C we find that the area between -2 and 2 under the normal curve is approximately .95 (.954 to be more exact). The remark after Example 6.21 shows that the same is true for the points $x_1 = \mu - 2\sigma$ and $x_2 = \mu + 2\sigma$ for an arbitrary normal distribution with mean μ and standard deviation σ. If we apply this result to our binomial distribution, we find that with probability approximately .95 the observed number of successes falls between $np - 2\sqrt{npq}$ and $np + 2\sqrt{npq}$. In our case $np = 50$ and $\sqrt{npq} = \sqrt{25} = 5$, so that it is reasonable to expect that the number of heads is somewhere between 40 and 60 (see Problem 12).

PROBLEMS

1 Find the area under the normal curve to the left of
   a. 1.55,
   b. -1.55,
   c. $-.86$,
   d. 4.25.

2 Find the area under the normal curve to the right of
   a. 1.11,
   b. $-.45$,
   c. 3.75,
   d. -3.75.

3 Find the area under the normal curve between
   a. 1.28 and 1.96;
   b. -1 and 2;
   c. -1.15 and 1.15.

4 Find the area under the normal curve outside the interval
   a. $-.85$ to .85;
   b. -2 to 2.

5 Find the value of z such that the area under the normal curve to the left of z is
   a. .719,
   b. .66,
   c. .008,
   d. .34.

6 Find the value of z such that the area under the normal curve to the right of z is

a. .30,

b. .78.

7 Find the value of z such that the area under the normal curve between −z and z is

a. .90,

b. .68,

c. .34.

8 For the normal distribution with mean 500 and standard deviation 100 find

a. the area to the left of 450;

b. the area to the right of 650;

c. the area between 550 and 650;

d. the numbers that contain the central 90% of the area;

e. the numbers that contain the central 95%;

f. the value x such that the area to the right of x equals .05;

g. the value x such that the area to the left of x equals .05.

9 Use the normal approximation to find the following binomial probabilities:

a. $P(k \geq 10)$ if $n = 25$ and $p = \frac{1}{2}$ (Compare your result with the exact probability obtained from Table A.);

b. $P(20 \leq k \leq 30)$ if $n = 80$ and $p = .30$;

c. $P(40 < k < 60)$ if $n = 125$ and $p = .35$.

10 A fair coin is tossed 400 times. Find the probability that the resulting number of heads is between 181 and 219, both limits included.

11 Use the normal approximation to show that the probability of observing exactly 50 heads and 50 tails in 100 tosses with a fair coin is approximately .08. Look up the exact probability in an appropriate table.

12 It is implied in the statement at the end of this chapter that if a fair coin is tossed 100 times, the probability is approximately .95 that the resulting number of heads (or tails) lies between 40 and 60. The statement does not specify whether the two limits are to be included or excluded. Use the normal approximation and/or a table of exact probabilities to find

a. $P(40 \leq k \leq 60)$,

b. $P(40 < k < 60)$.

Do you see now why the statement is left intentionally vague?

*13 Explain why the answers to Problem 4.13 imply that the binomial distribution has mean np and standard deviation $\sqrt{npq}$, the quantities we have used when approximating binomial probabilities by means of a normal distribution.

7

Estimation

41 In Chapter 5 we discussed the following type of problem: We plan to observe an experiment that consists in performing n independent trials, each with two possible outcomes, success and failure, and with constant known success probability p. What is the probability that we observe some given number of successes? For example, what is the probability that in 48 trials with success probability $\frac{1}{3}$ we observe 12 successes? This is a purely probabilistic problem.

Now let us consider a different setup. We have actually observed the results of 48 trials of the type described above, say, 12 successes and 36 failures. But we do not know the value of the success probability p that brought about this result. We want to use the available information to estimate the unknown value of p. This is a statistical problem.

In statistics a quantity like p that is characteristic of the given experiment, but whose true value is usually unknown, is called a *parameter*. The success probability p is a parameter of the binomial distribution. Other examples of parameters are the mean μ and the standard deviation σ of a normal distribution. In this chapter we discuss methods for estimating the parameter p of a binomial distribution.

A Point Estimate for p

42 In the above example most readers would presumably decide to use the observed relative frequency of success, $\frac{12}{48} = \frac{1}{4}$, as their estimate. In Chapter 1 in connection with the problem of estimating the number of taxis in a city we pointed out that there may be more than one possible estimate for a given unknown quantity.

54

The same is true of the success probability p. Estimates other than the one based on the relative frequency are possible. However in this book we shall consider only this particular estimate. Thus if in n independent trials with constant (unknown) success probability p we observe k successes and $n - k$ failures, we shall estimate p as the relative frequency of success, k/n.

How good is this estimate? It is implied in the frequency interpretation of probability that if the number of trials is sufficiently large, the relative frequency of success, and therefore our estimate of p, can be expected to differ arbitrarily little from the success probability p. The question is, how large is sufficiently large? There is no unique answer to this question. Whenever we generalize from sample information, we must be prepared for erroneous conclusions, unless we are satisfied with trivial statements. The best that we can hope for are statements that have a certain known probability of being correct. Thus we have to reformulate our question. How many trials are needed so that we can be reasonably sure that our estimate of the success probability p will not deviate from the true value p by more than some given positive quantity δ. Here "reasonably sure" means with a given high probability, for example, with probability .95. We can translate this into mathematical language by requiring that the following inequality hold with probability .95:

$$-\delta \le \frac{k}{n} - p \le +\delta.$$

If δ is small, as we certainly would want it to be, our requirement is satisfied only if the number of trials is large. However, in that case we can use the normal approximation in place of exact binomial probabilities. In particular we saw that approximately with probability .95 the number of successes k does not differ from the expected number of successes np by more than two standard deviations, or $2\sqrt{npq}$. In mathematical language, the following inequality holds approximately with probability .95,

$$-2\sqrt{npq} \le k - np \le 2\sqrt{npq}.$$

This second inequality is easily converted into the form of the first inequality by dividing all three parts by n. Then

$$-2\sqrt{\frac{pq}{n}} \le \frac{k}{n} - p \le 2\sqrt{\frac{pq}{n}}$$

and we see that

7.1
$$\delta = 2\sqrt{\frac{pq}{n}}$$

or

$$n = \frac{4pq}{\delta^2}.$$

The number of trials should be $4pq/\delta^2$, if we want to be 95% certain that our estimate does not differ from the true value p by more than δ. However this result still does not solve our problem.

Since we want to estimate p, the value of p is presumably unknown, yet our formula for determining n depends on p. This looks like a vicious circle. But let us not give up yet. The quantity that enters into the determination of n is the product pq. Here is a table giving the value of pq for various values of p:

TABLE 7.2

p	.5	.4 or .6	.3 or .7	.2 or .8	.1 or .9	.05 or .95
pq	.25	.24	.21	.16	.09	.0475

We see that pq is never larger than $.25 = \frac{1}{4}$ and progressively decreases as p deviates from $\frac{1}{2}$. Thus whatever the true value of p, we always have $4pq/\delta^2 \le 1/\delta^2$, so our number of trials need not exceed $1/\delta^2$ under any circumstances.

TABLE 7.3

δ	.10	.05	.02	.01
$n = 1/\delta^2$	100	400	2,500	10,000

Table 7.3 evaluates $1/\delta^2$ for some typical values of δ. We see that if we do not mind an error of .05, 400 trials are sufficient. But we need 10,000 trials if we want to be reasonably certain that our estimate is in error by at most .01.

Table 7.3 can also be used in reverse. Suppose that we have decided to observe the results of 100 trials and then estimate the unknown success probability p as the relative frequency k/n. Table 7.3 tells us that we can be reasonably sure that this estimate is not going to be in error by more than .10.

An interesting application of the preceding results is to public opinion polls such as those conducted before presidential elections. While the sampling scheme underlying such polls is considerably more complicated than the one we have discussed, our formula still can be applied to give some indication of a poll's accuracy. Suppose that our poll is based on 2500 interviews. (Actually, most polls use fewer.) Table 7.3 tells us that an estimate based on such a poll may be in error by as much as 2 percentage points. A candidate who gets only 48.5% of the votes of those participating in a poll may nevertheless be the choice of the majority of all voters.

Let us return briefly to Table 7.2. It is seen that for p between .3 and .7 the product pq is not much smaller than $\frac{1}{4}$. If p can be expected to be in this interval (and the statistician usually has such rough information), we essentially need $n = 1/\delta^2$ trials to produce a satisfactory estimate. On the other hand if p is close to 0 or 1, substantially fewer trials are sufficient. Thus for instance if we have reason to think that the true value of p is smaller than $\frac{1}{10}$ (or greater than $\frac{9}{10}$), it is sufficient to use only

$$n = 4 \times \frac{1}{10} \times \frac{9}{10} \times \frac{1}{\delta^2} = \frac{.36}{\delta^2}$$

trials, a saving of 64%.

43   When planning a statistical investigation the question often arises
of how many more observations we would need in order to double
our accuracy. Of course we have to explain what we mean by
doubling our accuracy. In the present case a very simple interpretation is possible. Since in estimating p we must be prepared to
commit a certain error, doubling our accuracy should be interpreted
as cutting this possible error in half. Since δ is a measure of the
possible error, this means cutting δ in half. It may seem that we
should be able to do this by taking twice as many observations.
But let us look again at our formula determining n as a function of
δ, $n = 4pq/\delta^2$. If we replace δ by $\delta/2$, we find

$$n = \frac{4pq}{\delta^2/4} = \frac{16pq}{\delta^2}.$$

The new n must be 4 times as large as the old n. Since in general it
takes time and money to provide additional observations, there
comes a point when costs of additional observations more than
outweigh any possible gain due to greater accuracy. This is a
typical example of the law of diminishing returns. If we have 100
observations, 300 additional observations double our accuracy.
But 300 additional observations do very little to increase our
accuracy if we already have 1000 observations.

The relationship that we have just discovered between the
number of observations and the resulting accuracy is not limited
to the problem of estimating the parameter p of a binomial
distribution. It arises over and over again in other statistical
problems. Increasing the accuracy of a statistical procedure by a
factor of 2 usually requires 4 times as many observations; tripling
the accuracy requires 9 times as many observations. And a ten-fold
increase in accuracy can be achieved only by a 100-fold increase
in the number of observations. Indeed such an increase in the
number of observations may well be accompanied by complications that would upset any expected gain in accuracy.

44   The estimate k/n of the success probability p is called a *point
estimate*. The reason for this name is rather obvious. The value k/n
represents a single point in the interval from 0 to 1 of all possible
p-values. In practical applications point estimates are used more
frequently than any other kind of statistical procedure. But point
estimates do have their limitations.

Consider an experiment where we observe 55 successes in 100
trials. Now suppose that in another experiment involving 10000
trials we observe 5500 successes. In both cases the point estimate
of p is .55, and there is no way of distinguishing the second
estimate from the first. Nevertheless, our earlier considerations
imply that the accuracy of the second estimate is 10 times as large
as the accuracy of the first estimate. Unless we specifically add a
statement that indicates in some way the degree of accuracy of a
given point estimate, we do not know whether the estimate is
highly reliable or does not even deserve being called an estimate.
In the past when you read a public opinion poll which stated that,

according to its findings, 55% of the people preferred proposition A to proposition B, did you ever wonder by how much this statement might be in error? Public opinion polls rarely, if ever, mention how accurate their estimates are.

Interval Estimates

45 Let us look again at Table 7.3 relating n and δ. We see from this table that a point estimate based on 100 observations may reasonably be expected to be in error by no more than .10, while the corresponding error of an estimate based on 10000 observations is only .01. Assume that both times the point estimate is .55. In the first case the true value of p would then be somewhere between $.55 - .10 = .45$ and $.55 + .10 = .65$. In the second case the corresponding two values are $.55 - .01 = .54$ and $.55 + .01 = .56$. We shall call such an interval of possible p-values a *confidence interval* for the parameter p. In our specific example we have the following picture:

n	k	point estimate	confidence interval
100	55	.55	$.45 \leq p \leq .65$
10000	5500	.55	$.54 \leq p \leq .56$

The much shorter length of the second interval clearly shows the superiority of the second estimate. Indeed the length of the second interval is $.56 - .54 = .02$, while that of the first interval is $.65 - .45 = .20$. The length of the first interval is 10 times that of the second interval, showing immediately the 10-fold increase in accuracy as we go from 100 to 10000 trials. Unless a problem of statistical estimation specifically requires a point estimate as an answer, interval estimates are usually preferable. We shall now investigate interval estimates in greater detail.

46 Let us start by indicating how the lower and upper endpoints of the appropriate interval, the *confidence limits*, are found in general. Suppose we have observed k successes in n trials. We first compute the point estimate k/n. The lower endpoint is found by subtracting δ from k/n, the upper endpoint by adding δ to k/n. Thus we have the confidence interval

7.4
$$\frac{k}{n} - \delta \leq p \leq \frac{k}{n} + \delta.$$

According to our earlier results, $\delta = 1/\sqrt{n}$ so that (7.4) becomes

7.5
$$\frac{k}{n} - \frac{1}{\sqrt{n}} \leq p \leq \frac{k}{n} + \frac{1}{\sqrt{n}}.$$

In particular if $k = 55$ and $n = 100$, we get our previous result.

This interval has the following interpretation. Theoretically the unknown success probability p may have any value between 0 and 1. However, on the basis of the available information, namely k successes in n trials, we are confident that the true value of p really lies in the shorter interval bounded by

$$\frac{k}{n} - \frac{1}{\sqrt{n}} \text{ and } \frac{k}{n} + \frac{1}{\sqrt{n}}.$$

As we have mentioned earlier, statisticians can never be 100% certain of their statements, only reasonably certain. So far, we have interpreted "reasonably certain" to mean with probability .95. Accordingly, the above confidence interval is said to have *confidence coefficient* .95. From a practical point of view, confidence intervals with confidence coefficient .95 contain the true but unknown success probability p about 95% of the time. On the other hand, 5% of the time the true value of p is not contained in the confidence interval computed on the basis of available sample information. Of course in any particular case the statistician does not know whether he is dealing with the first or the second situation.

47   We now want to discuss how to find a confidence interval for the parameter p that has confidence coefficient γ, where γ is arbitrary. We also want to investigate how a change in γ affects such a confidence interval.

We limit our discussion to confidence intervals of type (7.4). These intervals are bounded by lower and upper limits equal to

$$\frac{k}{n} - \delta \text{ and } \frac{k}{n} + \delta,$$

respectively. According to (7.1) we obtain a confidence interval with confidence coefficient .95 by setting $\delta = 2\sqrt{pq/n}$. The .95 confidence coefficient results from the fact that the area under the normal curve between -2 and $+2$ equals (approximately) .95. A confidence interval with confidence coefficient γ results from setting $\delta = z\sqrt{pq/n}$, where z is read from Table C corresponding to γ. We then have the interval

7.6
$$\frac{k}{n} - z\sqrt{\frac{pq}{n}} \le p \le \frac{k}{n} + z\sqrt{\frac{pq}{n}}.$$

We can now see the effect of changing the confidence coefficient on the confidence interval. According to (7.6), the length of the confidence interval is $2z\sqrt{pq/n}$. Thus the length of the interval increases as z increases. Since z increases together with the confidence coefficient γ, a larger confidence coefficient results in a longer interval. Therefore, when deciding on a confidence coefficient we should balance the advantages of a safer statement against the disadvantages of a longer interval.

Since p and q are unknown, the two limits in (7.6) can not be evaluated numerically. As in Section 42 we may decide to replace $\sqrt{pq}$ by its maximum value $\frac{1}{2}$. This results in the confidence interval

7.7
$$\frac{k}{n} - \frac{z}{2}\sqrt{\frac{1}{n}} \leq p \leq \frac{k}{n} + \frac{z}{2}\sqrt{\frac{1}{n}}.$$

For $z = 2$, (7.7) reduces to (7.5).

Unless p happens to have the value $\frac{1}{2}$, the interval (7.7) is actually longer than is necessary to achieve the *nominal* confidence coefficient corresponding to the value z used in computing the limits (7.7). Looking at it a different way, we can say that the true confidence coefficient associated with the interval (7.7) is really higher than we claim. Such an interval is said to be *conservative*. As long as the true value of p is close to $\frac{1}{2}$, there is little difference between the true and nominal confidence coefficients. However if p is close to 0 or 1, the difference may be substantial and the conservative interval may not represent a very desirable solution. As an alternative statisticians often use the estimate k/n of p in computing the value of $\sqrt{pq}$. With this the confidence interval (7.6) becomes

7.8
$$\frac{k}{n} - z\sqrt{\frac{(k/n)(1 - (k/n))}{n}} \leq p \leq \frac{k}{n} + z\sqrt{\frac{(k/n)(1 - (k/n))}{n}}.$$

EXAMPLE 7.9
Find a confidence interval with confidence coefficient .90 for the success probability p, when in 400 trials we observe 40 successes. Corresponding to $\gamma = .90$ in Table C we find the value $z = 1.645$. Since
$$\frac{k}{n} = \frac{40}{400} = \frac{1}{10} = .10$$
we have
$$z\sqrt{\frac{(k/n)(1 - (k/n))}{n}} = 1.645\sqrt{\frac{(.10)(.90)}{400}} = .025$$
so that the confidence interval (7.8) becomes
$$.10 - .025 \leq p \leq .10 + .025$$
or
$$.075 \leq p \leq .125.$$
To determine the conservative interval (7.7) we compute
$$\frac{z}{2}\sqrt{\frac{1}{n}} = \frac{1.645}{2}\sqrt{\frac{1}{400}} = .041.$$
Thus we have the interval
$$.10 - .041 \leq p \leq .10 + .041$$
or
$$.059 \leq p \leq .141.$$

Comparing this with the previous result $.075 \leq p \leq .125$, we see that in this case the conservative confidence interval is unnecessarily wide.

The need for finding a confidence interval for the parameter p of a binomial distribution arises so frequently in practice that mathematicians and statisticians have expended a great deal of thought and effort on the problem. The methods we have discussed have the advantage of simplicity and are sufficiently accurate for many practical purposes. However, there exist tables and graphs that speed up the process of finding confidence limits and are more accurate than our methods, especially for small values of n when the normal approximation to the binomial distribution is not satisfactory. We shall see an example of such a table in Section 54.

PROBLEMS

1 You have observed the results of 100 trials in order to find a point estimate of the success probability p. You then decide that you really want an estimate that is three times as accurate as your present estimate. How many additional trials should you perform in order to attain this accuracy?

2 You want a point estimate of the success probability p that (with probability .95) does not deviate from the true value p by more than .025. How many trials should you perform in order to obtain such an estimate?

3 If in Problem 2 you suspect that p is at least .8, how many trials would you perform?

4 A statistician has told you that a confidence interval of length .10 for the success probability p requires 400 trials. You can perform only 100 trials. How long is your confidence interval going to be? What decision is implicit in your answer?

5 Find confidence intervals for the success probability p corresponding to the following information:
   a. $n = 100, k = 60, \gamma = .95$;
   b. $n = 100, k = 10, \gamma = .95$;
   c. $n = 100, k = 40, \gamma = .95$;
   d. $n = 100, k = 40, \gamma = .90$;
   e. $n = 100, k = 40, \gamma = .99$;
   f. $n = 400, k = 80, \gamma = .99$;
   g. $n = 25, k = 10, \gamma = .90$.

6 In Problem 1 of Chapter 2 you were asked to roll a die 120 times. Define success as "the die shows 5 or 6." Find a confidence interval for the success probability p. Does your interval contain the value $p = \frac{1}{3}$?

7 Problem 2, Chapter 2, was concerned with 180 imaginary rolls of two dice. Define success as "the two dice show a total of 10 or more." Find a confidence interval for the success probability p. Does the interval contain the value $p = \frac{1}{6}$?

8 Repeat Problem 7, using the actual data called for in Problem 3, Chapter 2.

9 Consider the data in Table 3.5. Define success as "the second digit equals the first digit." Find a confidence interval for the success probability p. Does the confidence interval contain the value $\frac{1}{3}$?

10 Repeat Problem 9, using the data in Table 3.6.

11 Take 50 samples of 100 random digits each. For each sample, count the number k of odd digits and find a confidence interval with confidence coefficient .90 using Formula 7.7. How many of your intervals contain the correct success probability $p = \frac{1}{2}$? It is instructive to represent your results graphically. For this purpose each interval is marked off on a separate horizontal line as follows:

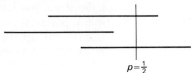

$$p = \frac{1}{2}$$

In the present example we happen to know that the true success probability is $\frac{1}{2}$. Thus we know which intervals cover the true success probability and which intervals do not. But in general the true success probability p is unknown. We then do not know where p is in relation to the computed confidence limits. All we know is that the long-run frequency with which confidence intervals cover the true parameter value equals the chosen confidence coefficient.

8

Tests of Hypotheses: The Null Hypothesis

Basic Ideas

48 In the preceding chapter we were interested in finding an estimate of the parameter p of a binomial distribution. Now we want to discuss a somewhat different problem. We may think that the success probability p associated with an experiment (for example, the mental random digit guesses discussed in Section 16) has some definite value, like $p = \frac{1}{3}$. We should then like to know whether or not our assessment of the value of p can be supported by experimental evidence. Using statistical terminology, we want to *test the hypothesis* that p equals $\frac{1}{3}$.

A *statistical hypothesis* is a statement about the distribution of observable quantities. A *test* of a statistical hypothesis is a procedure that permits us to *accept* or *reject* the hypothesis under consideration on the basis of available experimental data. Our interest in developing procedures for testing statistical hypotheses stems from the fact that our decision to accept or to reject a given hypothesis usually has an important bearing on our future course of action.

49 The question we have to answer is, how can we establish the correctness or incorrectness of a statistical hypothesis? A mathematical proposition is shown to be true or false by logical reasoning. But we are not dealing with a mathematical proposition. We are dealing with events subject to chance fluctuations. Logical reasoning is unable to prove or disprove a statistical hypothesis. We need some other approach. The reader may remember cases where a single counter-example disproved the validity of a mathematical proposition. Perhaps some similar approach works in statistics. Let us look at an example.

As the curtain rises on the play by Tom Stoppard, *Rosencrantz and Guildenstern Are Dead,* we see Hamlet's two boyhood friends engaged in an animated game of flipping coins. We are given to understand that the coins have turned up heads some eighty times in succession and keep turning up heads. Philosophizing about life in general and probability in particular, our two heros agree that some deep unknown forces are at play. As they prepare to obey the summons to appear at court, both are overcome by a feeling of impending doom.

Let us see what we can learn from this incident. Actually a less dramatic version reveals more clearly the particular argument we are trying to illustrate. Suppose we watch a person flip a coin 10 times, and each time the coin falls heads. We are very likely to think that some cheating may be going on. Let us try to establish a parallel between this train of thought and the use of a counter-example to disprove a mathematical proposition:

set up a mathematical proposition (example: all prime numbers are odd)	set up a statistical hypothesis (example: coin tosses are fair)
give a counter-example (2 is both prime and even)	observe sample results (10 heads in 10 tosses)
conclusion: mathematical proposition is false	conclusion: we have doubts about the correctness of our statistical hypothesis

As far as the mathematical proposition is concerned, it is false beyond any doubt. As far as the correctness of the statistical hypothesis is concerned, we cannot be quite as definite. But we might argue as follows. While it is possible to get 10 heads in 10 tosses of a fair coin, the probability of such an event is less than one in a thousand ($(\frac{1}{2})^{10} = \frac{1}{1024}$, to be exact). Since we actually did observe 10 heads in 10 tosses, it seems more plausible to believe that there has been some cheating than to believe that the coin tosses were fair. In future dealings with the person who flipped the coins, we had better be on guard!

What general approach does our example suggest to the problem of testing a statistical hypothesis? More specifically, how can we find out whether or not the parameter p associated with certain trials has a given value p_0? It is customary in statistics to indicate a hypothetical parameter value by using a subscript zero. The hypothesis being tested is very often called the *null hypothesis* and symbolized by H_0. In particular, in our case we speak of a test of the hypothesis $H_0: p = p_0$.

The test of H_0 is to be based on the number k of successes that we observe in an experiment involving n independent trials. If H_0 is correct, we expect k to be in the neighborhood of np_0. Thus, a value of k close to np_0 does not raise doubts about the correctness of the hypothesis. On the other hand, the more k deviates from np_0, the more untenable the null hypothesis becomes. But where shall we draw the dividing line between acceptance and rejection of H_0? Before we try to answer this question, we need to introduce some further terminology.

50 A value k that leads to rejection of H_0 is called a *rejection value*. All rejection values taken together form the *rejection* or *critical region*. To determine a test procedure for a statistical hypothesis means to determine a critical region. Whenever our experiment results in a value k in the critical region, we reject the hypothesis being tested; if the value k is not in the critical region, we accept the hypothesis. In constructing a critical region for carrying out a test we shall always use the following intuitively reasonable principle:

Suppose a critical region contains a value k. Then if k is greater than np_0, the critical region also contains the values $k + 1$, $k + 2, \ldots, n$; if k is less than np_0, it also contains the values $k - 1$, $k - 2, \ldots, 0$. This principle considerably limits the critical regions we might possibly use, but we need one further criterion to help us make a choice.

The reason why we may hesitate to declare 10 heads in 10 coin tosses as dishonest is the fact that it is possible for 10 honest coin tosses to produce 10 heads. We could be accusing an honest person of cheating. Similarly in our general procedure, even if H_0 is true, there is always the possibility that the number of successes resulting from an experiment corresponds to one of the values in the critical region. We would then decide to reject the hypothesis being tested, and this rejection would be incorrect. Clearly we should like this to happen only rarely. This suggests the following criterion for selecting a critical region. If the hypothesis is true, the probability of observing a critical number of successes should be small. This probability of falsely rejecting the hypothesis being tested when it is actually true is called the *significance level* of the test and denoted by the Greek letter α.

The choice of a specific value for α is not strictly a statistical problem. Considerations of the possible consequences of a false rejection—in our case, the unwarranted accusation of cheating—should have considerable influence on the choice of α. Practicing statisticians often use routine levels like .05 or .01 or levels as close to these as possible. But these levels are not prescribed by theory and often simply reflect a desire on the part of the statistician to do as others do without asking why. From the practical point of view, use of a .05-level implies that in approximately 1 out of every 20 cases in which the hypothesis being tested is actually correct, we decide to act as if it were false.

Let us return briefly to the example where we observe 10 heads in 10 tosses of a coin. While it is true that such an occurrence has only a very small probability when tossing a fair coin, this by itself is not sufficient reason to reject the null hypothesis. Equally important is the fact that a much more plausible explanation exists, namely cheating. In general when we test the hypothesis $p = p_0$, we do so because we know that the true value of p may differ from p_0, and if so, we should like to find out. The possibility that an *alternative* to the null hypothesis may be true raises problems that we shall study in greater detail in Chapter 9. At this moment we simply observe that the possibility that we reject a true null hypothesis is only one side of the problem. There is also the possibility that we do not reject a false null hypothesis that ought to be rejected. Our coin tosser may be cheating, but he may also be clever enough not to make it obvious. The question is, will we catch him nevertheless? As we shall see in Chapter 9, our choice of α has an important bearing on the probability that we shall do so. The smaller a value of α we choose, the more difficult we make it for us to catch a cheater. For the remainder of this chapter however, we shall pay attention mainly to the first type of error, the error that consists of rejecting a null hypothesis that happens to be correct.

Finding a Critical Region

51 We again take the case

$$H_0: p = \tfrac{1}{2}$$

where the test is to be based on $n = 10$ trials. As before, we may think of the coin tossing activities of a person suspected of cheating. We want to find possible critical regions for testing our hypothesis. One additional requirement that we have not yet mentioned seems appropriate in this case. If we should decide to accuse our coin tosser of cheating, if he produces, say, 9 heads and 1 tail, we should also do so if he produces 1 head and 9 tails. After all, a person who is able to manipulate the coin in such a way that it produces more heads than tails should also be able to achieve the opposite result as well. And we cannot predict what he is going to do.

Our criteria suggest consideration of only the following three critical regions (see Problem 1):

   (i) $k = (0 \text{ or } 10)$　　　　　$\alpha = .002$

   (ii) $k = (0, 1, 9, 10)$　　　　$\alpha = .022$

   (iii) $k = (0, 1, 2, 8, 9, 10)$　　$\alpha = .110$

The α-levels follow from the binomial probability table for $n = 10$ and $p = \tfrac{1}{2}$. Thus for region (ii), $\alpha = b(0) + b(1) + b(9) + b(10) = .022$. If we decide to reject the hypothesis $p = \tfrac{1}{2}$

whenever we observe 0, 1, 9, or 10 heads in 10 tosses of the coin, we have roughly 2 chances in 100 of declaring the coin tosses unfair when in reality $p = \frac{1}{2}$.

Many statisticians would feel that a significance level of .110 is too large. But it is better not to take such a rigid attitude. The example shows one thing very clearly. With 10 observations we cannot hope to recognize anything but the crudest violations of the null hypothesis. For example, the second critical region requires at least a 9 to 1 split between heads and tails before we can reject the null hypothesis. Such a result is not very likely to occur unless the true probability of heads is very close to zero or very close to one. Thus we have practically no chance of finding out that the coin is not fair unless the bias is very strong.

The situation gets progressively better as we take more and more observations. Table 8.1 lists critical regions that have α-levels

TABLE
8.1

Critical Regions for Testing $H_0: p = \frac{1}{2}$ with Significance Level $\alpha = .05$

n	critical regions	split
100	$(0, \ldots, 40, 60, \ldots, 100)$	40 : 60
1000	$(0, \ldots, 469, 531, \ldots, 1000)$	47 : 53
10000	$(0, \ldots, 4901, 5099, \ldots, 10000)$	49 : 51

as close to .05 as we can get (see Problem 8). Thus for $n = 10000$, a split of at least 49 to 51 among heads and tails assures rejection of the null hypothesis. Such a split is very likely to occur when the coin is only slightly biased in favor of heads or tails.

52 When testing the hypothesis $p = \frac{1}{2}$, symmetry considerations suggest that we put the value $k = 0$ in the critical region along with $k = n$, the value $k = 1$ along with $k = n - 1$, and so on. But suppose that we want to test the hypothesis $p = .4$ on the basis of 10 observations. If the hypothesis is correct, we expect to observe in the neighborhood of 4 successes. We should decide to reject the hypothesis only if the actually observed number of successes differs sufficiently from 4. However, this time earlier symmetry considerations do not apply, since $b(0)$ does not equal $b(10)$, $b(1)$ does not equal $b(9)$, and so on. In such a case it is usually appropriate to build up the two "tails" of the critical region subject to the following two conditions:

   (i) The probability of all k-values in the critical region should correspond to the chosen significance level α.

   (ii) The two tails should as nearly as possible have equal probabilities.

In the example of testing the hypothesis $p = .4$, if we want approximately $\alpha = .10$, we can use the critical region $k = (0, 1, 7,$

8, 9, 10). For this critical region the lower tail has probability $\alpha_1 = b(0) + b(1) = .006 + .040 = .046$, while the upper tail has probability $\alpha_2 = b(7) + b(8) + b(9) + b(10) = .042 + .011 + .002 + .000 = .055$, so that the true significance level is $\alpha = \alpha_1 + \alpha_2 = .101$. When the number n of trials is small, it is at times impossible to equalize the two tail probabilities satisfactorily. However, equalization becomes less and less of a problem as n increases.

Descriptive Levels

53   The following approach avoids setting up a formal critical region and at the same time is often more informative. After we have performed our experiment, that is, after we have observed k successes in n trials, we ask ourselves the following question. If the hypothesis is correct and if we should repeat the experiment, how likely is it that we should again observe as extreme a result as the one that we did observe? This probability has been called the *descriptive level of significance*. If the descriptive level is smaller than or equal to the significance level α that we have chosen for the test, we reject the hypothesis. Otherwise we accept it. An example will explain the procedure in greater detail.

EXAMPLE 8.2   We want to test the hypothesis $p = .4$. If we observe 8 successes in 10 trials, can we reject our hypothesis at significance level .10? In order to find the descriptive level associated with this result, we need the probability of 8 or more successes in 10 trials with success probability .4. According to Table A this probability is $b(8) + b(9) + b(10) = .011 + .002 + .000 = .013$. However this is not the descriptive level since it only refers to one of the two tails of an appropriate critical region. In order to take care of the other tail, we simply double this probability. The descriptive level for our experiment is then .026, considerably smaller than the significance level .10 that we were willing to tolerate. Thus we reject the null hypothesis.

In reporting the results of a statistical analysis, it is usually preferable to state that the descriptive level is .026 rather than simply to say that the null hypothesis is rejected at significance level .10. In this way a reader who feels that a different significance level is more appropriate has the possibility to decide for himself whether or not the null hypothesis under investigation should be rejected at the significance level he has chosen for his own use.

EXAMPLE 8.3   Consider the same problem as in Example 8.2, except that we observe 2 successes. Now we compute $b(2) + b(1) + b(0) = .121 + .040 + .006 = .167$. Even without doubling, as we should in order to find the descriptive level for our test, we note that this probability is greater than .10. We therefore decide that there is no reason to reject the given hypothesis.

EXAMPLE 8.4

We want to test the hypothesis $p = .8$. What can we say if we observe 105 successes in 144 trials? Since n is large, we use the normal approximation to compute relevant binomial probabilities. We find $\mu = np = (144)(.8) = 115.2$; $\sigma = \sqrt{npq} = \sqrt{(144)(.8)(.2)} = 4.8$. To find the descriptive level associated with our experimental result, we consider $P(k \leq 105)$. According to Chapter 6 we need the z-value corresponding to $x = 105 + \frac{1}{2}$, $z = (105 + \frac{1}{2} - 115.2)/4.8 = -2.02$. The lower tail probability associated with this z-value is .022, so that the descriptive level of our result is approximately .044.

Tests of Hypotheses and Confidence Intervals

54

In Chapter 3 we found a confidence interval for the parameter T of the taxi problem by testing the hypothesis $T = T_0$ for various values T_0. The confidence interval consisted of all values T_0 such that the hypothesis $T = T_0$ did not lead to rejection of the hypothesis being tested. The confidence interval for T consisted of *acceptable* T-values or, reversing the argument, a value $T = T_0$ was acceptable if it lay in the confidence interval for T.

This mutual relationship between tests of hypotheses on the one hand and confidence intervals on the other hand holds for parameters other than T as well. We shall often make use of this property as we study new statistical procedures. Now we look at two examples involving the parameter p of a binomial distribution.

EXAMPLE 8.5

We return to the data for the experiment in Section 16 in which students were asked to mentally draw a sequence of two pingpong balls each labeled 1, 2, or 3. When we first saw the data in Table 3.5, even though we were relying on intuition, it seemed fairly evident that the number of students who selected the same digit twice was considerably lower than could be expected in a similar experiment with real pingpong balls. Now we have the tools for a more thorough investigation. For real pingpong balls the probability of getting the same digit twice in a row is $\frac{1}{3}$. Suppose we let p be the probability that a person participating in our experiment puts down the same digit twice in succession. We can then test the hypothesis that $p = \frac{1}{3}$, the same as for draws of real pingpong balls. Rather than find a critical region and simply reject or accept the hypothesis $p = \frac{1}{3}$, it is more interesting to find a confidence interval for p. In order to be really confident of our conclusions, let us make it a 99% confidence interval. The only bit of information we need is the number of successes. This we get from Table 3.5. There are 62 double-ones, 54 double-twos, and 58 double-threes. In all then, the quantity k equals $62 + 54 + 58 = 174$. When we substitute this value in Formula 7.8 using $n = 900$ and $z = 2.576$, we find the confidence interval

$$.16 \leq p \leq .23.$$

We note that the value $p = \frac{1}{3}$ is not in the interval. Thus we may say that on the basis of available information, the value $p = \frac{1}{3}$ does not deserve our confidence. This of course is just another way of saying that we should reject the hypothesis $p = \frac{1}{3}$. What about the significance level of the corresponding test? Since we are 99% confident in our confidence interval statement, there is only one chance in 100 that we do not include the correct value p in the interval. In particular if $p = \frac{1}{3}$ is the correct value, then there is only one chance in 100 that the value $\frac{1}{3}$ is not included in the confidence interval. The significance level α is 1 in 100 or .01.

The specific relationship between the confidence coefficient of a confidence interval and the significance level of the corresponding test that we have just found in Example 8.5 can be generalized. If γ is the confidence coefficient and α, the significance level, then we have

$$\alpha = 1 - \gamma.$$

We have observed that Table C can be interpreted as a table of z-values that produce confidence intervals with confidence coefficients γ. The second column of Table C gives the significance level $\alpha'' = 1 - \gamma$ of the related test procedure. As we shall see in subsequent chapters, the third column is also related to tests of hypotheses.

EXAMPLE 8.6   Earlier in this chapter we discussed tests of the hypotheses $p = \frac{1}{2}$ based on the results of $n = 10$ trials. The critical region with significance level .05 or less turned out to be $k = (0, 1, 9, 10)$. Thus at the .05 significance level the hypothesis $p = \frac{1}{2}$ is rejected if in 10 trials we observe 0, 1, 9, or 10 successes. In all other cases the hypothesis $p = \frac{1}{2}$ cannot be rejected at the chosen significance level. The same (and more) information is revealed by Table 8.7, which for $k = 0, 1, \ldots, 10$ lists confidence intervals for p with confidence coefficients at least .95 corresponding to k successes in 10 trials. The intervals for $k = 0, 1, 9, 10$ successes do not contain the value $p = \frac{1}{2}$, while the remaining intervals do contain the value $p = \frac{1}{2}$.

TABLE 8.7

Confidence Intervals for p; $n = 10$, $\gamma \geq .95$

k	interval	k	interval
0	$0 \leq p \leq .31$	10	$.69 \leq p \leq 1$
1	$0 \leq p \leq .45$	9	$.55 \leq p \leq 1$
2	$.03 \leq p \leq .56$	8	$.44 \leq p \leq .97$
3	$.07 \leq p \leq .65$	7	$.35 \leq p \leq .93$
4	$.12 \leq p \leq .74$	6	$.26 \leq p \leq .88$
5	$.19 \leq p \leq .81$	5	$.19 \leq p \leq .81$

A confidence interval is usually much more informative than a test of a hypothesis. Assume, for example, that we have observed

$k = 6$ successes in 10 trials. According to Example 8.6, the hypothesis $p = \frac{1}{2}$ is accepted by both the test based on the critical region $k = (0, 1, 9, 10)$ and the appropriate confidence interval, $.26 \leq p \leq .88$. However the test provides information about the parameter value $p = \frac{1}{2}$ only. The confidence interval on the other hand shows that parameter values like $\frac{1}{3}$ and $\frac{3}{4}$ are also acceptable when we observe 6 successes in 10 trials. Indeed the length of the confidence interval reveals that the result of 10 trials sheds very little light about the true value of p.

Sampling from a Finite Population

55   The binomial model assumes that the outcomes of successive trials are independent. We have seen in Section 18 that when we sample from a finite population without replacement the results of successive selections are not independent. It is then incorrect to base exact probability computations on the binomial model. However if the sample size is large, approximate probabilities can be obtained using the normal approximation of Section 39 by multiplying the binomial standard deviation $\sqrt{npq}$ by the factor $\sqrt{(N - n)/(N - 1)}$, where N is the size of the population.

56   *1969 Draft Lottery*   Many people have pointed out that the 1969 draft lottery assigns more low draft numbers to men whose birthdays fall late in the year than to men whose birthdays fall early in the year. We may ask whether the observed discrepancy can be explained in terms of random fluctuations or whether it is indicative of possibly insufficient shuffling of the capsules that contained the draft dates. There are quite a few statistical procedures that can be brought to bear on this question. Here we shall use a very simple approach. (See, however, also Problem 9, Chapter 16.)

Let us call a date with draft number 183 or lower "vulnerable" and a date with draft number 184 or higher "non-vulnerable". In the 1969 draft, 73 of the 182 days from January to June turned out to be vulnerable while 109 turned out to be non-vulnerable. What is the probability of observing this large a discrepancy between the number of vulnerable and non-vulnerable dates in a random lottery?

Vulnerable dates result from the first 183 draws in the draft lottery. The probability that any one of these draws produces a January-through-June date is $\frac{182}{366}$. We have then 183 trials with "success" probability $p = \frac{182}{366}$ and can rephrase our question as: What is the probability of observing 73 or fewer successes in 183 trials with success probability $\frac{182}{366}$? Since sampling in the draft lottery is without replacement from a population of $N = 366$ dates, the normal approximation should be based on the mean $\mu = np = 183 \times \frac{182}{366} = 91$ and the standard deviation

$$\sigma = \sqrt{npq}\,\sqrt{\frac{N-n}{N-1}} = \sqrt{183\left(\frac{182}{366}\right)\left(\frac{184}{366}\right)}\,\sqrt{\frac{366-183}{366-1}} = 4.79.$$

We approximate the probability of observing 73 or fewer vulnerable dates by the area under the normal curve to the left of

$$z = \frac{73 + \frac{1}{2} - 91}{4.79} = -3.65.$$

This area is .0001. As in the case of computing the descriptive level associated with an observed test result, this probability should be doubled to take care of the possibility of observing too many vulnerable January-through-June dates. Even the doubled probability is so small that it raises strong doubts about the complete random selection of draft dates in the 1969 draft lottery.

The above analysis requires one further comment. We have treated the problem as a test of the hypothesis $p = \frac{182}{366}$. Had we decided to carry out such a test *before* the draft lottery actually took place, our conclusions would be fully justified at the stated descriptive level. However what actually happened was that we used the given data to test a hypothesis that came to our attention exactly because of the large observed discrepancy between reality and expectation. In general, statistical hypotheses suggested by a set of data should not be tested by using the same data. They should be tested on the basis of newly obtained data.

Actually, in the draft example, it is possible to look at the test that we have performed as one of a large number of tests that should be performed routinely to check the overall randomness of the draft order.

PROBLEMS

1 Find the significance level of the test of the hypothesis $p = \frac{1}{2}$, $n = 10$, where the critical region is given as $k = (0, 1, 2, 3, 7, 8, 9, 10)$.

2 Set up possible critical regions to test each of the following hypotheses when a sample of the indicated size is available. State the significance level of each test.
   a. $p_0 = .50$, $n = 20$.
   b. $p_0 = .50$, $n = 25$.
   c. $p_0 = .60$, $n = 10$.
   d. $p_0 = .20$, $n = 20$.

3 You are testing the hypothesis $p = .4$ using 25 observations. You decide to reject the hypothesis if either $k \le 5$ or $k \ge 15$, where k is the observed number of successes. Find the significance level of your test using
   a. exact binomial probabilities,
   b. the normal approximation.

4 You have found the confidence interval $.68 \le p \le .73$ having confidence coefficient .95. Consider the following two statements.
   (i) The hypothesis $p = \frac{2}{3}$ is accepted at significance level .05.
   (ii) The hypothesis $p = \frac{3}{4}$ is accepted at significance level .05.
   Which, if any, of these two statements is correct?

5 Reformulate Problems 6–10 of Chapter 7 as tests of appropriate hypotheses. In each case determine whether your hypothesis should be accepted or rejected. State the significance level of your tests.

6 Use the data in Table 11.1. Test the hypothesis that the probability that a person has brown eyes is $\frac{1}{2}$.

7 Use the data in Table 11.1. Test the hypothesis that the probability that a person has dark hair is $\frac{1}{2}$.

8 Verify that the critical regions in Table 8.1 for sample sizes 100, 1000, 10000 have significance level .05.

9 The confidence intervals for $n = 10$ in Table 8.7 are based on exact computations (beyond the scope of this course). In order to show that the normal approximation may or may not give satisfactory results for n as small as 10, find 95% confidence intervals using Formula 7.8 when
   a. $k = 5$,
   b. $k = 8$.

9

Tests of Hypotheses: Alternative Hypothesis

Extra-Sensory Perception

57 Mrs. A claims that she has extra-sensory perception. Mr. B doubts her claim and offers her a one hundred dollar reward if she can convince him of her abilities. An experiment is arranged. Mr. B is going to use 26 cards, a red and a black ace, a red and a black king, and so on, down to a red and a black two. He is going to hold up one pair of cards at a time, one card in his left hand and the other card in his right hand, in such a way that Mrs. A can see only the backs of the cards. Nevertheless, she is going to tell which card is red and which is black. The experiment is performed and Mrs. A makes 8 correct and 5 incorrect calls. Satisfied with her performance, she demands her reward. But Mr. B does not agree. He contends that Mrs. A has been guessing and that her results can easily be explained in terms of chance. They decide that the problem calls for statistical analysis, and this is where we come in.

The problem most certainly calls for statistical analysis, but it would have been much better if a statistician had been consulted before rather than after the experiment. As we shall see, the whole experiment is most inconclusive. Mrs. A and Mr. B are left pre tt much where they started—Mrs. A convinced of her supernatural powers and Mr. B as skeptical as ever. But we are getting ahead of our analysis.

Let p be the probability that Mrs. A indicates the correct arrangement for a single pair of cards. We have to decide whether p is equal to $\frac{1}{2}$ or greater than $\frac{1}{2}$. The value $p = \frac{1}{2}$ represents guessing, pure and simple, while a value of p greater than $\frac{1}{2}$ indicates something in addition to guessing, whatever it might be. Investigations of extra-sensory perception experiments have sometimes brought to light incredibly poor experimental controls.

We set up a test of the null hypothesis $H_0: p = \frac{1}{2}$ and arrange the test in such a way that rejection of this null hypothesis implies acceptance of Mrs. A's claim. The reader may wonder at this moment why we are discussing this example. After all, the greater part of the preceding chapter was concerned with the same hypothesis $p = \frac{1}{2}$. And 13 observations rather than 10 observations do not make such a difference. This is perfectly true, but there is another important difference. In the coin tossing example we decided to reject the null hypothesis if we observed too many or too few heads, indicating that the true value of p might be greater than $\frac{1}{2}$ or less than $\frac{1}{2}$. Indeed we used this argument to convince ourselves that we should consider only symmetric critical regions. This time the situation is different. We want to reject the null hypothesis only if there is a sufficiently large number of correct identifications indicating a value of p that is greater than $\frac{1}{2}$. This time our critical region should be *one-sided*, containing only large values of k, where k is of course the number of correct identifications that Mrs. A manages to make. Let us see what critical regions are available to us and what the associated significance levels are. The table of the binomial distribution with $n = 13$ gives the required information. This information is summarized in the upper part of Table 9.1.

TABLE
9.1

Test Probabilities, $n = 13$

r	7	8	9	10	11	12	13
significance level α for critical region $k \geq r$	.500	.291	.134	.047	.012	.002	.000
rejection probability $1 - \beta$ when $p = \frac{2}{3}$	.896	.759	.552	.322	.139	.039	.005
failure probability β when $p = \frac{2}{3}$	.104	.241	.448	.678	.861	.961	.995

The most extreme critical region available to us contains only the value $k = 13$. Its significance level α is $b(13)$, which according to the table is zero to three decimal places. This means that $\alpha < .0005 = \frac{1}{2000}$. The next critical region contains both 12 and 13 and its α-value is $b(12) + b(13) = .002$.

If we have our heart set on a test with significance level approximately .05, we have to use the critical region containing the values $k = 10, 11, 12,$ and 13. Since Mrs. A identified only 8 out of 13 pairs, there is no reason to reject the hypothesis $p = \frac{1}{2}$, the hypothesis that says that Mrs. A has simply been guessing. Even if we should have been willing to use a significance level as large as .13, the available information still does not suggest rejection of the null hypothesis. This is of course the reason why Mr. B was quite unimpressed by the performance of Mrs. A and why he felt justified in withholding the reward.

However there is another side to the picture. Let us assume, just for the sake of argument, that Mrs. A is not merely guessing. What are her chances of collecting the promised reward? To have something more specific to talk about, let us assume that she has probability $p = \frac{2}{3}$ of making a correct identification. Such a value of p does not necessarily imply clairvoyant powers. Mr. B may have a tendency to alternate red and black according to a regular pattern. If Mrs. A learns from observation, she should be able to achieve a rather high score. Mr. B can prevent such a possibility by using a mechanical randomizing procedure—like the toss of a coin or the use of random digits—to determine for each pair of cards which card goes on the right and which on the left. But at this moment we are not concerned with this aspect of the matter. We only want to know the probability that Mrs. A makes a number of correct identifications sufficiently high to reject the hypothesis of mere guessing. Now we need probabilities based on $p = \frac{2}{3}$ rather than $p = \frac{1}{2}$. These probabilities are given in the middle row of Table 9.1 (labeled "rejection probability").

If we use the test with significance level approximately .05, Mrs. A has less than one chance in three of showing that she is not merely guessing. This probability goes down still further if we use a more stringent critical region. The significance of these probabilities will perhaps become clearer if we give a reinterpretation of our problem. Suppose that an instructor announces that at the end of his course every student will have to take an examination consisting of 13 true-and-false questions. In order to get a passing grade for the course, a student would have to answer at least 10 questions correctly. Otherwise he would fail the course. Those are exactly the requirements Mrs. A has to meet in order to collect the reward, if we are to apply a test with significance level .05.

Of course, students who know the course material well should have no difficulty answering at least 10 out of 13 questions correctly. But what about a marginal student? To be more specific, let us take a student who knows the answers to approximately $\frac{2}{3}$ of the type of questions that are customarily asked on an examination of this type. Such a student would seem to be entitled to at least a low pass. But what does our table tell us? Such a student has less than one chance in three of getting a passing grade for the course. Perhaps the issue becomes even clearer when we look at the last line in Table 9.1. There we find the probabilities that our marginal student fails the course. Alternatively, these are the probabilities that Mrs. A is unable to convince Mr. B of her special abilities.

Type 1 and Type 2 Errors

58   The time has come to look more carefully at the problem of testing statistical hypotheses. Everything seemed very simple as long as we concentrated only on the null hypothesis. We ran into difficulties when we started to ask questions about what happens if the null hypothesis is not true, if, as the statistician says, an *alternative hypothesis* is true.

When carrying out a test of a statistical hypothesis we can go wrong in two completely different ways depending on whether the null hypothesis that we are testing happens to be true or false. When the null hypothesis is true and when on the basis of sample evidence we decide to reject it, we commit what is customarily called a *type 1 error*. When the null hypothesis is false and when on the basis of sample evidence we do not decide to reject it, we commit what is customarily called a *type 2 error*. The various possibilities that may occur when testing a hypothesis are represented schematically in Table 9.2.

TABLE 9.2

Correct and Incorrect Decisions in Hypothesis Testing

action taken as result of test	*unknown true situation*	
	null hypothesis true $(p = \frac{1}{2})$	alternative hypothesis true $(p = \frac{2}{3})$
accept null hypothesis (give no reward)	correct decision probability: $1 - \alpha$	type 2 error probability: β
reject null hypothesis (give reward)	type 1 error probability: α	correct decision probability: $1 - \beta$

It should be fairly clear now why we ran into difficulties. In setting up a test, we only worried about the type 1 error, giving Mrs. A a reward that she did not deserve. We insisted that the associated probability α should be small. At no time did we consider the possibility of a type 2 error, withholding a deserved reward, and its probability. Table 9.1 shows that the probabilities of these two errors are inversely related. As we try to decrease one probability, we increase the other, at least as long as we keep other aspects of the experiment unchanged.

Clearly in the extra-sensory perception experiment we cannot have both probabilities arbitrarily small at the same time. We have to compromise somewhere. It often happens that one error is much more serious than the other. In such a case it is reasonable to make the probability of the more serious error smaller than that of the less serious error. But in our example it seems reasonable to make the two error probabilities as similar as possible. This occurs for the critical region $k \geq 8$. Of course neither error probability is very satisfactory. The probability of giving an unearned reward is .29, and the probability of withholding an earned reward is .24.

The only real way out of the difficulty is to change the experimental setup. Thirteen observations are simply not enough to discriminate satisfactorily between the two probabilities $\frac{1}{2}$ and $\frac{2}{3}$. This is exactly what a statistician could have pointed out before the start of the whole experiment. What is more, he could have determined how many trials should have been performed in order to reduce both probabilities to reasonable levels.

59 In order to see how the error probabilities decrease as the number of observations increases, let us look at an example involving 50 observations. In particular, let us think of a true-and-false examination containing 50 questions.

This time the parameter p is the probability that a student answers a question correctly. Since we deal only with true-and-false questions, even a completely unprepared student can achieve a value of $p = \frac{1}{2}$ simply by tossing a coin and, say, answering T (for true) every time the coin falls heads and F (for false) when it falls tails. We want to answer two questions. What is the probability that a totally unprepared student receives a passing grade? And what is the probability that a prepared student receives a failing grade?

We interpret "totally unprepared" to mean that a student essentially guesses whether to mark a given statement true or false by using either real or imagined coin tosses. In terms of the parameter p, an unprepared student is one for whom $p = \frac{1}{2}$. A prepared student is one for whom p is greater than $\frac{1}{2}$. More specifically we shall assume, at least for the moment, that $p = \frac{2}{3}$, indicating of course a fairly marginal student. Our procedure can then be viewed as a test of the

$$\text{null hypothesis: } p = \frac{1}{2}$$

against the

$$\text{alternative hypothesis: } p = \frac{2}{3}$$

using 50 observations as the basis of our test.

Before we compute the probabilities of type 1 and type 2 errors, it is helpful to state specifically the nature of these errors. We commit a type 1 error when we reject a true null hypothesis. In our example a type 1 error is committed when the instructor gives a passing grade to a student who deserves to be failed. A type 2 error occurs if the null hypothesis is accepted when in reality the alternative hypothesis happens to be true. In our example a type 2 error is committed when on the basis of his examination score our student receives a failing grade, even though his general knowledge warrants a passing grade. We mentioned earlier that the seriousness of the given errors should determine to some extent the α and β probabilities that we are willing to tolerate. The present example shows that different persons may have different ideas about the relative seriousness of the two errors. A student taking the examination would hardly worry about the type 1 error, receiving a passing grade, when he really should have failed the course. But he would certainly object to receiving a failing grade when he felt that he deserved a passing grade. The instructor on the other hand may feel that both kinds of errors are equally undesirable.

We shall now compute actual probabilities. Let k be the number of questions answered correctly. A student who guesses can expect to answer approximately 25 questions correctly by chance alone. A student who has some knowledge of the subject matter should be able to do better. Thus, sufficiently large values of k indicate that the null hypothesis should be rejected. The critical region for our test consists of sufficiently large values of k, that is, all values of k that are greater than or equal to some value r.

In Table 9.3 we find both α and β probabilities for various choices of r. We notice the error probabilities are considerably lower than those in Table 9.1. But error probabilities of .10 and .13

TABLE 9.3

Type 1 and Type 2 Error Probabilities
$H_0: p = \frac{1}{2}, H_1: p = \frac{2}{3}, n = 50$

r	29	30	31	32
$\alpha = P(k \geq r)$ when $p = \frac{1}{2}$	.161	.101	.059	.032
$\beta = P(k < r)$ when $p = \frac{2}{3}$	.076	.126	.196	.287

(corresponding to $r = 30$) are still somewhat high, suggesting that even 50 questions on a true-and-false examination is not sufficiently discriminatory.

The Power Curve

60   Some readers may have wondered why we have concentrated on the value $p = \frac{2}{3}$ in discussing type 2 errors. There is really no particular reason. Any other value of p greater than $\frac{1}{2}$ could have served equally well. Indeed, when evaluating the properties of a test procedure statisticians like to look at several different p-values. Thus in evaluating the test that rejects the null hypothesis $p = \frac{1}{2}$ if the observed number of successes is at least 30 ($r = 30$), we should consider the kind of information contained in Table 9.4.

TABLE 9.4

Rejection Probabilities
$p_0 = \frac{1}{2}, n = 50$, critical region: $k \geq 30$

true p-value	.45	.50	.55	.60	.65	.70
rejection probability	.024	.101 (=α)	.286	.561	.814	.952

Such information can be plotted on a graph by marking p along the horizontal axis and the rejection probability along the vertical axis as in Figure 9.5. The resulting curve is called the *power curve*

FIGURE 9.5

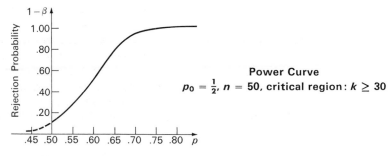

Power Curve
$p_0 = \frac{1}{2}, n = 50$, critical region: $k \geq 30$

of the test. Power curves enable the statistician to evaluate the effectiveness of a given test procedure and to compare two or more competing test procedures.

In the present problem where we want to decide whether or not a student is to receive a passing grade, it has little meaning to ask what happens when $p = .45$ or some other value smaller than $\frac{1}{2}$. One possible interpretation of such a p-value is that the student is not guessing but that he has actually acquired some incorrect information that reduces his overall performance. Table 9.4 reveals that a student whose p-value is .45 has probability .024 of receiving a passing grade. Information like this is important in situations that are mathematically identical with the present problem but involve different practical circumstances. We consider the following (oversimplified) example.

61   A standard medical treatment has been observed to produce undesirable side effects in 50% of the patients to whom it is given. It is claimed that a modification of the treatment, which is considerably more complicated than the original treatment, would decrease this percentage without changing the effectiveness of the treatment itself. The question that has to be answered is whether doctors should use the modified treatment in the future, in place of the old treatment.

Here we have a situation where it is necessary to choose between two possible courses of action, recommending or not recommending the use of the modified treatment. This is a typical situation that calls for an appropriate test of a statistical hypothesis, where we decide on one course of action if the test should lead to acceptance of the hypothesis being tested and a different kind of action if the test should reject the hypothesis. In the present situation, since the modified treatment is still experimental, it is not known what percentage of patients to whom the modified treatment is given will stay free of side effects. Let us denote this unknown proportion by p. (With this notation, a success is a patient who undergoes the modified treatment and does not develop any side effects.) If p should be $\frac{1}{2}$ or less, use of the modified treatment in place of the old treatment is highly undesirable. Not only is the modified treatment more complicated, but it also represents no improvement over the old treatment and actually may be worse. On the other hand, the more p exceeds $\frac{1}{2}$, the more desirable it becomes to replace the old treatment by the modified treatment. The problem of choosing between the two possible courses of action can be decided statistically by setting up a test of the hypothesis $p = \frac{1}{2}$ and choosing a critical region that contains only large k-values. Acceptance of the hypothesis $p = \frac{1}{2}$ leads to the recommendation that the modified treatment should not be used in the future; rejection leads to the opposite recommendation. If in particular we carry out an experiment involving 50 patients and decide to recommend use of the modified treatment if at least 30 of these 50 patients do not show any side effects, then the information in Table 9.4 (and Figure 9.5) allows the following reinterpretation. If $p = \frac{1}{2}$, the same as for the old treatment, the probability is .101, or 1 chance in 10, that we are going to recommend use of the more complicated treatment modification even though the modification does not change the incidence of side effects. There is 1 chance in 40 (probability .024 to be exact) that we are going to recommend the treatment modification even though only 45% of all patients are going to stay free of side effects. On the other hand if p should equal .70, there are about 19 chances in 20 (probability .952 to be exact) that we are going to recommend introduction of the modified treatment. With this kind of information at their disposal, medical authorities can then decide whether an experiment involving 50 patients provides sufficient information to make the kind of decision that is required, or whether possibly a larger (or smaller) experiment is called for.

Sample Size

62 The problem of determining an appropriate sample size is often solved as follows. In addition to the hypothetical value p_0 and the associated type 1 error probability α, we select another value p_1 such that if the true success probability is p_1, we should like the probability of a type 2 error to be some small quantity β. In other words, p_0 and p_1 are selected in such a way that we are very eager to do one thing (stick with the old treatment in our medical example) if $p = p_0$ or less and do something else (switch to the modified treatment) if $p = p_1$ or larger. Any test of the type discussed in this chapter that satisfies the given two requirements has a power curve similar to the one illustrated in Figure 9.6.

FIGURE 9.6

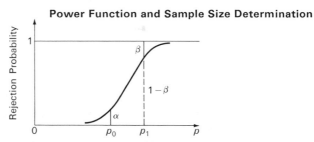

Power Function and Sample Size Determination

If $p \leq p_0$, the probability of deciding to act as if $p = p_1$ is less than or equal to α. If $p \geq p_1$, the probability of deciding to act as if $p = p_0$ is less than or equal to β. It can then be shown that the number n of observations required for such a test is approximately

$$n = \frac{(z_0\sqrt{p_0 q_0} + z_1\sqrt{p_1 q_1})^2}{(p_1 - p_0)^2}$$

where z_0 is read from Table C corresponding to $\alpha' = \alpha$ and z_1 is read from Table C corresponding to $\alpha' = \beta$.

EXAMPLE
9.7

If $p_0 = \frac{1}{2}, \alpha = .10, p_1 = .7, \beta = .05$, then

$$n = \frac{(1.282\sqrt{\frac{1}{2} \times \frac{1}{2}} + 1.645\sqrt{.7 \times .3})^2}{(.7 - .5)^2} = 49.$$

(Note that the test underlying Table 9.4 has $p_0 = \frac{1}{2}, \alpha = .101$, $p_1 = .7$, $\beta = 1 - .952 = .048$, and $n = 50$, in very close agreement with our present result.)

In practice experimenters often like to have tests that are able to discriminate with small error probabilities between values p_0 and p_1 that differ very little. Such discrimination is possible only on the basis of a large number of observations. If the experimenter does not have the facilities to conduct such a large scale experiment, he must decide whether he can scale down his requirements or has to abandon his experiment altogether. It is certainly better to be forewarned than to find out after performing the experiment that the information obtained is not good enough to make a reliable decision, as in the experiment conducted to prove or disprove Mrs. A's powers of clairvoyance.

Summary Remarks

63 Before going on to tests of other hypotheses, it is helpful to summarize certain general ideas. When we test a null hypothesis H_0 like $p = p_0$, we are often really seeking guidance about which of two possible actions to take, an action A that is appropriate if $p = p_0$ or an action R that is appropriate when the true value of p differs from the hypothetical parameter value p_0 in some way or other. When we set up a test of the hypothesis H_0 then, we should keep in mind that acceptance of H_0 has the practical implication of taking action A; rejection of H_0 has the practical implication of taking action R. When we talk of type 1 and type 2 errors (or α-errors and β-errors, as they are sometimes called), we are really talking about the error of taking action R when we should have taken action A and vice versa. This realization throws new light on the meaning of the power curve. Whatever the true value p, the power curve indicates the probability with which we are going to take action R when we use the given test procedure.

Since we reject H_0 and take action R whenever k falls in the critical region, the critical region should consist of k-values that are indicative of the alternative parameter values for which action R is appropriate. It is good practice to state along with the null hypothesis H_0 the alternative H_1 against which the test is to be effective. Thus for the coin problem of Chapter 8, we test the null hypothesis $H_0: p = \frac{1}{2}$ against the alternative $H_1: p \neq \frac{1}{2}$. This alternative requires a two-tailed critical region. In the problems of this chapter we test the hypothesis H_0: $p = \frac{1}{2}$ against the one-sided alternative $p > \frac{1}{2}$, suggesting an upper tail critical region. In such a problem it often is more logical to write the null hypothesis as $p \leq \frac{1}{2}$, since we usually are even more interested in taking action A when $p < \frac{1}{2}$ than when $p = \frac{1}{2}$, as we saw in the medical example. The power curve in Figure 9.5 shows that the critical region consisting of large k-values achieves this purpose: if $p = \frac{1}{2}$, the probability of taking action A is $1 - \alpha$, and as p becomes smaller, this probability increases towards 1.

In general there are three different test situations:

(i) H_0: $p = p_0$ against H_1: $p \neq p_0$, requiring a two-tailed critical region

(ii) H_0: $p \leq p_0$ against H_1: $p > p_0$, requiring a critical region consisting of large k-values

(iii) H_0: $p \geq p_0$ against H_1: $p < p_0$, requiring a critical region consisting of small k-values

Our medical problem could have been formulated as a type (iii) problem by defining success as the occurrence, rather than non-occurrence, of side effects.

It should be clear from the discussion in this chapter, that a knowledge of the power of a test is of great importance in the proper conduct of an experiment. Power computations are relatively simple for the type of problems that we have been discussing in Chapters 8 and 9. However, this is no longer the case for the problems to be discussed in subsequent chapters. While we shall still mention the general principles that we have developed in this chapter, we shall not try to carry out further power computations. Such computations have to be reserved for more advanced treatments.

PROBLEMS

1 For each of the following hypotheses set up possible critical regions and determine the associated significance levels. In all cases, assume samples of size 20.

a. H_0: $p = .50$ against H_1: $p > .50$.
b. H_0: $p = .50$ against H_1: $p < .50$.
c. H_0: $p = .30$ against H_1: $p > .30$.
d. H_0: $p = .30$ against H_1: $p < .30$.

2 Use the normal approximation and/or tables of exact binomial probabilities to check the entries in
a. Table 9.3,
b. Table 9.4.

3 You are testing the hypothesis $p = .30$ against the alternative $p < .30$ and have decided to reject the hypothesis if in 25 trials you observe fewer than 5 successes.
a. What is the significance level of your test?
b. Find the probability of committing a type 2 error if the true value of p is .20. Repeat for $p = .10$, $p = .05$.

c. What is the probability of accepting the hypothesis if $p = .40$? By accepting the hypothesis in this case, have you committed an error (if so, which error) or have you made the correct decision?

d. Use the probabilities computed for parts (a), (b), and (c) of this question to construct the power curve for the test.

4 You want to test the hypothesis $p = .2$ against the alternative $p > .2$ using 13 trials. Find the appropriate critical region corresponding to a significance level .10.

5 You are testing the hypothesis $p = .4$ on the basis of 25 trials. You decide to reject the hypothesis if either $k \leq 5$ or $k \geq 15$, where k is the observed number of successes.

a. Against what alternative is this test appropriate?

b. What is the significance level of the test?

c. If the true value of p is .2, what is the probability that the test accepts the hypothesis being tested?

d. If the event in (c) occurs, which error if any has been committed?

6 Consider the test that rejects the hypothesis $p = .6$ if in 20 trials at most 9 successes are observed.

a. Against what alternative is the test appropriate?

b. What is the significance level of the test?

7 You are testing the hypothesis $p = .1$ against the alternative $p > .1$ on the basis of 100 trials. As your critical region you have chosen $k \geq 15$. According to the normal approximation, what is the significance level of this test?

8 Past experience shows that if a certain machine is property adjusted, 5% of the items turned out by the machine are defective. Each day the first 25 items produced by the machine are inspected for defects. If 3 or fewer defects are found, production is continued without interruption. If 4 or more items are found to be defective, production is interrupted and an engineer is asked to adjust the machine. After adjustments have been made, production is resumed. This procedure can be viewed as a test of the hypothesis $p = .05$ against the alternative $p > .05$, p being the probability that the machine turns out a defective item. Using test terminology, the engineer is asked to make adjustments only when the hypothesis is rejected.

a. What is the critical region of the test?

b. What is the significance level of the test?

c. What is the practical meaning of a type 1 error?

d. What is the practical meaning of a type 2 error?

e. If the machine has gone out of adjustment and produces 20% defectives, what is the probability of a type 2 error?

f. The foreman would like to have at most 1 chance in 100 of stopping production needlessly for machine adjustments. What is the smallest number of items out of 25 that have to be found defective to achieve this purpose?

9 Find at least 7 points on the power curve of the test of the hypothesis $p = \frac{1}{2}$ using the critical region $k = (0, 1, 2, 8, 9, 10)$. The total number of trials is 10.

10 a. How many true-and-false questions should be included on an examination so that a student who is guessing has 1 chance in 20 of receiving a

passing grade while a student who knows the answers to 60% of the type of questions appearing on the examination has 1 chance in 10 of failing?

b. How many of these questions would a student have to answer correctly in order to obtain a passing grade?

c. What is the probability that a student who knows the answers to 70% of the type of questions passes the examination?

11 The data in Example 12.6 represent yearly incomes of 20 families in a certain community. Each family had been instructed to report the family income to the nearest 100 dollars. Do you think that the respondents heeded the instruction? Justify your answer.

10

Chi-Square Tests

64   We have seen how to analyze the results of experiments in which each trial has only two possible outcomes. Now we want to take up the case when a trial has more than two possible outcomes. Here are a few examples.

In a table of random digits there are 10 possibilities for each position, namely the digits 0, 1, . . . , 9. When rolling two dice, the sum of the number of points showing can range from 2 to 12, a total of 11 possibilities. At many colleges a student gets one of 5 possible grades, A, B, C, D, or F, in a course he takes for credit. A geneticist is interested in the various types of offspring that result from the crossing of certain parental traits.

It may sometimes be appropriate to reduce the number of categories to two and then apply one of the statistical analyses discussed earlier. Thus in the case of a student's grades, we may simply indicate whether the student passes or fails. But often such a simplification ignores important details. A passing grade may be given for outstanding work as well as for work that is barely satisfactory. We need methods for analyzing data resulting from trials with more than two outcomes or categories. The following notation is useful in discussing such methods.

number of categories: m
total number of observations: n
number of observations in ith category: $O_i, i = 1, \ldots, m$
probability associated with ith category: $p_i, i = 1, \ldots, m$

Clearly we must have $O_1 + \cdots + O_m = n$ and $p_1 + \cdots + p_m = 1$.

In general the true probabilities $p_1, \ldots, p_m$ are unknown, but we have certain ideas about what values these probabilities might or should have. Thus for rolls of two *fair* dice it is easy to show that

TABLE
10.1

Probabilities for Rolls of Two Fair Dice

number of points	2	3	4	5	6	7	8	9	10	11	12
probability	$\frac{1}{36}$	$\frac{2}{36}$	$\frac{3}{36}$	$\frac{4}{36}$	$\frac{5}{36}$	$\frac{6}{36}$	$\frac{5}{36}$	$\frac{4}{36}$	$\frac{3}{36}$	$\frac{2}{36}$	$\frac{1}{36}$

the probabilities associated with various numbers of points are those given in Table 10.1. But now suppose that we have two of the decorative dice some people like to keep on a shelf or desk. Are they fair? Usually such dice do not roll very well, so it is not at all certain whether the probabilities in Table 10.1 apply to such dice. What we can do is to roll the two dice a large number of times and find out how often the number of points adds up to 2, 3, and so on. We can then compare these actually observed frequencies with frequencies expected for ideal dice. If there is reasonable agreement between the two sets of numbers, we would decide that the dice are fair. If the agreement is bad, we had better use the dice for decorative purposes only. One question arises immediately. How are we going to decide whether the agreement between actually observed frequencies and theoretical frequencies is reasonable? We want to find an answer to just this question, but first let us look at some actual data.

Table 10.2 contains the results of 180 rolls of two dice together with theoretical frequencies denoted by E_i, as well as some other information that we shall need presently. The theoretical frequencies are easily computed. For example, for fair dice, the probability of getting 2 points is $\frac{1}{36}$. In 180 rolls the expected number of 2's is $180 \times \frac{1}{36} = 5$. The other expected frequencies are computed in a similar fashion.

TABLE
10.2

Results of 180 Rolls of Two Dice

number of points	p_i	O_i	E_i	$O_i - E_i$	$(O_i - E_i)^2/E_i$
2	$\frac{1}{36}$	6	5	+1	.20
3	$\frac{2}{36}$	7	10	−3	.90
4	$\frac{3}{36}$	11	15	−4	1.07
5	$\frac{4}{36}$	19	20	−1	.05
6	$\frac{5}{36}$	26	25	+1	.04
7	$\frac{6}{36}$	28	30	−2	.13
8	$\frac{5}{36}$	32	25	+7	1.96
9	$\frac{4}{36}$	19	20	−1	.05
10	$\frac{3}{36}$	14	15	−1	.07
11	$\frac{2}{36}$	16	10	+6	3.60
12	$\frac{1}{36}$	2	5	−3	1.80
totals	1	180	180	0	9.87

Looking at the two sets of numbers labeled O_i and E_i, should we say that the agreement is reasonably good or not? Some observed frequencies agree with the corresponding expected frequencies surprisingly well. Others seem to disagree rather badly. What is the overall picture? Can the observed deviations be explained in terms of chance fluctuations that occur in all such experiments, or is there something more seriously wrong? We need an objective criterion for making our decision. Such a criterion is developed in the next section.

The Chi-Square Statistic

65 We are considering the following problem. We want to test a hypothesis about m probabilities $p_1, \ldots, p_m$ associated with m categories. For example, in the dice problem we have $m = 11$ and we want to test the hypothesis that

$$p_1 = \tfrac{1}{36}, p_2 = \tfrac{2}{36}, \ldots, p_{11} = \tfrac{1}{36}.$$

As data we have certain observed frequencies $O_1, O_2, \ldots, O_m$. These we can compare with corresponding theoretical or expected frequencies, which we call $E_1, E_2, \ldots, E_m$. The expected frequencies are computed from the probabilities $p_1, p_2, \ldots, p_m$ by multiplying these probabilities by the total number of observations n: $E_1 = np_1, \ldots, E_m = np_m$. It follows immediately that

10.3
$$
\begin{aligned}
E_1 + \cdots + E_m &= np_1 + \cdots + np_m \\
&= n(p_1 + \cdots + p_m) \\
&= n \\
&= O_1 + \cdots + O_m.
\end{aligned}
$$

Our problem is to find a criterion for comparing the observed frequencies $O_1, O_2, \ldots, O_m$ with the corresponding expected frequencies $E_1, E_2, \ldots, E_m$.

It should be clear by now that we are dealing with a generalization of the problem of testing a hypothesis about the parameter p of a binomial distribution. In that case we simply compared the observed number of successes, which we called k, with the expected number of successes np. If k differed too much from np we rejected the hypothesis being tested; otherwise we accepted it. How large this difference had to be before we could reject the hypothesis depended on the probability α of committing a type 1 error that we were willing to tolerate. In Chapters 8 and 9 we obtained the necessary information from tables of the binomial distribution or, when the number n of trials was large, by rejecting the given hypothesis if $z = (k - np)/\sqrt{npq}$ was either smaller than $-z_\gamma$ or greater than z_γ, where z_γ is the value read from Table C corresponding to $\gamma = 1 - \alpha$. Equivalently, for large n, we reject the hypothesis whenever $z^2 = (k - np)^2/npq$ is greater than z_γ^2.

Let us rewrite this latter test criterion using the notation of the present chapter. For two categories

$$E_1 = np, E_2 = nq, O_1 = k, O_2 = n - k,$$

so that

10.4
$$z^2 = \frac{(k - np)^2}{npq}$$

$$= (k - np)^2 \left[\frac{1}{np} + \frac{1}{nq} \right]$$

$$= (O_1 - E_1)^2 \left[\frac{1}{E_1} + \frac{1}{E_2} \right]$$

$$= \frac{(O_1 - E_1)^2}{E_1} + \frac{(O_2 - E_2)^2}{E_2}.$$

The last step follows from the fact that $O_1 + O_2 = E_1 + E_2$, and therefore, $O_2 - E_2 = -(O_1 - E_1)$.

The right side of (10.4) suggests that for m categories we use the following measure of discrepancy, usually denoted by χ^2 (chi-square):

10.5
$$\chi^2 = \frac{(O_1 - E_1)^2}{E_1} + \cdots + \frac{(O_m - E_m)^2}{E_m}.$$

An alternative formula for computing χ^2 is sometimes useful (see Problem 9):

10.6
$$\chi^2 = \frac{O_1^2}{E_1} + \cdots + \frac{O_m^2}{E_m} - n.$$

The test that consists in rejecting the hypothesis being tested when the value of χ^2 is sufficiently large is known as the *chi-square test*. This is one of the best-known tests in all of statistics.

66 The reason for rejecting our hypothesis for *large* values of χ^2 is as follows. If our hypothesis is correct, we should expect to find reasonably good agreement between the observed frequencies O_i and the theoretical frequencies E_i, resulting in a small value of χ^2, small that is, relative to the number of comparisons involved. On the other hand, if the hypothesis is not correct, we are using at least some incorrect values for the probabilities p_i when computing the corresponding theoretical frequencies E_i. This should result in some large discrepancies between the O_i and the E_i, discrepancies that produce a relatively large value of χ^2. We then choose a critical region consisting of sufficiently large values of χ^2. This still leaves the following question unanswered: For a given significance level α, where do we draw the dividing line between sufficiently small values suggesting acceptance and sufficiently large values suggesting rejection?

Our problem is simple when $m = 2$. In that case, χ^2 and z^2 are one and the same, and we know how to find a critical region based on z and therefore, z^2. However when the number of categories is greater than 2, a new table is required, a table of the so-called *chi-square distribution*. Table D is such a table.

Before we can describe this new table, it is necessary to introduce a new term. Associated with every chi-square test is a quantity called the number of *degrees of freedom*. For our test this quantity is simply $m - 1$, one less than the number of categories involved. However for other tests like the tests of association that we are going to discuss in the next chapter, the number of degrees of freedom is more complicated. An intuitive explanation for the number of degrees of freedom in the present test is as follows. In order to find the value of χ^2, we have to compute the m expected frequencies $E_1 = np_1, \ldots, E_m = np_m$. However, since by (10.3), $E_1 + \cdots + E_{m-1} + E_m = n$, we have $E_m = n - (E_1 + \cdots + E_{m-1})$, so theoretically it is sufficient to compute $E_1, \ldots, E_{m-1}$ according to the formula $E_i = np_i$ and then obtain E_m by subtraction. The important point is that E_m cannot be arbitrary. It is completely determined by the number of observations n and the remaining $m - 1$ theoretical frequencies $E_1, E_2, \ldots, E_{m-1}$. We therefore say that we have only $m - 1$ degrees of freedom.

We shall use the abbreviation *df* to denote the number of degrees of freedom. In Table D degrees of freedom are listed in the left-hand margin. In order to test our hypothesis at significance level α, we use the row in Table D for $m - 1$ degrees of freedom and find the tabulated value in the column corresponding to the upper tail probability α. If the computed value of χ^2 exceeds the tabulated value, we reject the hypothesis.

Let us return now to the dice example. From the last two columns in our table we find that chi-square for our data equals 9.87. We have $11 - 1 = 10$ degrees of freedom. If we want to test the hypothesis of fair dice at significance level .10, we find that the critical region extends beyond 16.0. Thus we accept our hypothesis. The observed deviations from expectation may be explained in terms of sampling fluctuations. Our table of the chi-square distribution tells us that in similar experiments with fair dice we can expect to observe a value of chi-square that is smaller than 9.87 about 55% of the time, while about 45% of the time we can expect a larger value.

We observed in the case of two categories that the test procedure based on z^2 as given by (10.4) requires that the number n of trials be sufficiently large. The same requirement is necessary for the more general χ^2-test. This raises the question of how large n should be in practice before we feel justified in using the χ^2-procedure. This is a rather complicated problem that cannot be answered satisfactorily in a few words. However the following somewhat overly conservative rule can provide guidance. The χ^2-procedure can be used satisfactorily if each theoretical frequency E_i is at least 5.

Two Examples

67   *Grading on a Curve*   In a course in elementary statistics attended by 283 students, the instructor gave the following grades: 25 A's, 77 B's, 114 C's, 48 D's, and 19 F's. We want to find out whether this instructor "grades on a curve." By "grading on a curve" we mean that the grades given out have a distribution that approximates a normal distribution.

Some time back we used areas under suitable normal distributions to approximate binomial probabilities. Areas under normal distributions can be used to describe many other phenomena as well. In particular, when grading on a curve the proportion of A, B, C, D, and F grades should correspond to areas under a normal distribution with suitable mean μ and standard deviation σ, as in Figure 10.7.

FIGURE 10.7

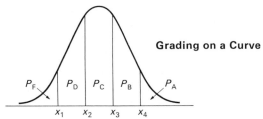

Grading on a Curve

Here x_1, x_2, x_3, x_4 are suitable dividing points and, for example, p_C refers to the proportion of C grades. If we choose $x_1 = \mu - \frac{3}{2}\sigma$, $x_2 = \mu - \frac{1}{2}\sigma$, $x_3 = \mu + \frac{1}{2}\sigma$, and $x_4 = \mu + \frac{3}{2}\sigma$, as is often done, we find $p_A = p_F = .07$, $p_B = p_D = .24$, and $p_C = .38$. We can then test the hypothesis that our 283 grades have come from a population of grades corresponding to these probabilities or proportions. We find theoretical frequencies $E_A = E_F = 283 \times .07 = 19.8$; $E_B = E_D = 283 \times .24 = 67.9$; $E_C = 283 \times .38 = 107.6$. Using expression (10.6) for a change, we find

$$\chi^2 = \frac{25^2 + 19^2}{19.8} + \frac{77^2 + 48^2}{67.9} + \frac{114^2}{107.6} - 283 = 8.83$$

with 4 degrees of freedom. This value of χ^2 is not significant at the .05 significance level, but it is significant at the .10 level. A comparison of the actual and theoretical grade frequencies reveals that our instructor has been rather lenient in assigning grades.

68   *How Not to Cheat*   According to our discussion, only sufficiently large values of χ^2 lead us to reject the hypothesis we are testing. But sometimes sufficiently small values of χ^2 also contain a message. Suppose you are asked to test a certain die for fairness. More specifically, suppose you are asked to roll the die 6000 times and note how often it falls 1, 2, 3, 4, 5, or 6. Rather than actually perform such a dreary experiment, some of you may be tempted to invent experimental results. The data in Table 10.8 certainly look reasonable for a fair die.

TABLE 10.8

Results of 6000 Imaginary Rolls of a Die

result of roll	1	2	3	4	5	6
number of observations	991	1005	1015	994	1007	988

The resulting value of χ^2 is .560. This chi-square value is so small that there is only 1 chance in 100 that it may have arisen in a real experiment with a fair die. As always, statistical evidence is not conclusive. Such data may come out of a real experiment. But they look very suspicious. Some data are just too good to be true. It takes considerable experience to fake data convincingly.

PROBLEMS

1 Suggest an appropriate hypothesis for the experiment in Problem 1 of Chapter 2 and use the data you obtained to test the hypothesis. Repeat for Problems 2–4 of Chapter 2.

2 Consider the data in Table 3.5.
   a. Test the hypothesis that a student writes down the digits 1, 2, or 3 as his first digit with probability $\frac{1}{3}$ each.
   b. Test the hypothesis that a student writes down each of the nine possible pairs (1, 1), (1, 2), . . . , (3, 3) with probability $\frac{1}{9}$.

3 Repeat Problem 2 using the data in Table 3.6. Compare with results of Problem 2 and explain any possible differences.

4 In a famous experiment the geneticist Gregor Mendel found that a sample from the second generation of seeds resulting from crossing yellow round peas and green wrinkled peas could be classified as follows:

yellow and round	315
yellow and wrinkled	101
green and round	108
green and wrinkled	32

According to the theory proposed by Mendel, the corresponding probabilities are $\frac{9}{16}, \frac{3}{16}, \frac{3}{16},$ and $\frac{1}{16}$, respectively. Does the experiment bear out these theoretical probabilities?

5 When 140 fatal traffic accidents were classified according to the day of the week on which the accident had occurred, the following table was obtained:

day	M	T	W	Th	F	Sa	S
number of fatal accidents	15	12	16	14	19	30	34

Are these data in agreement with the hypothesis that a fatal traffic accident is just as likely to occur on any day of the week?

6 Justify the probabilities in Table 10.1.

7 Consider the "Grading on a Curve" example (Section 67).
   a. Check the correctness of the probabilities for the various letter grades when grading on a curve.

b. Check the value of x^2 by using the other method of computation.

8 Check the value of x^2 for the "How Not to Cheat" example (Section 68).

9 Show that (10.6) is equivalent to (10.5). (*Hint:* Expand $(O_i - E_i)^2$ and simplify using (10.3).)

11

Tests of Independence and Homogeneity

69 So far we have always considered a single characteristic in classifying a given observation, for example, the grade of a student in a certain course, the number of points on a die, the number of 5's in tosses of 3 dice. However, there are situations where we observe more than one characteristic for every test subject in our sample. Thus, for each of a group of people we may note both the eye color and the hair color, say, blue, brown, and grey for eye color, and dark and light for hair color. We can then simply note how many people in the group have blue eyes, how many have brown eyes, and how many have grey eyes. Or we can note how many have dark hair and how many have light hair. However, if this is all we do, we might just as well have taken two separate groups of people, one to study the distribution of eye color, the other to study the distribution of hair color. An advantage of measuring more than one characteristic for each individual is to get information on possible interrelationships among these characteristics. Are hair and eye color related in any way? Is it more likely that a person with light hair has blue eyes than a person with dark hair? Or is the proportion of blue-eyed persons the same among dark- and light-haired individuals? These are the kinds of questions we want to answer.

First of all we have to organize our data. A two-way table, called a *contingency table*, is very natural. For our problem such a table might look like Table 11.1. The observations provide the numbers for the six cells in the center. Thus, in the upper left-hand corner we enter the number of persons in our sample who have light hair and brown eyes, in the lower left-hand corner, how many persons have dark hair and brown eyes, and so on. In addition, it is convenient to have totals for each row and each column. For example, the total for the first row, namely 40, tells us how many persons in our sample have light hair regardless of eye color. Since

TABLE
11.1

		brown	blue	grey	totals
hair	light	13	18	9	40
color	dark	37	12	11	60
	totals	50	30	20	100

eye color (spanning header over brown, blue, grey)

the totals are written in the margins, they are often referred to as marginal frequencies or totals.

A Test of Independence

70 The purpose of our investigation is to find out whether there exists a relationship or association between hair color and eye color. In order to get an answer, we specifically set up the null hypothesis that no such association exists. Only if an appropriate test rejects this hypothesis are we going to conclude that a relationship actually exists. But what test procedure should we use?

Let us look at our data. Our basic information consists of six frequencies, the numbers of persons in the six groups corresponding to all possible combinations of eye and hair color: brown eyes and light hair, blue eyes and light hair, and so on. These six observed frequencies remind us of the chi-square test of Chapter 10. But so far we do not have any expected frequencies with which we can compare the observed frequencies. In the preceding chapter we had hypothetical probabilities for each category under consideration. We computed the expected frequencies simply by multiplying these probabilities by the number of observations in the sample. This time we do not have such probabilities, at least not in explicit form. However, our null hypothesis states that eye and hair color are unrelated. Using the technical term that we introduced in Chapter 3, our hypothesis states that the two characteristics under investigation are independent of each other. We have to compute expected frequencies that reflect this assumption of independence.

Let, for example, B denote the event that a person has blue eyes and L, the event that he has light hair. Events B and L are independent if $P(B \text{ and } L) = P(B)P(L)$. If E'_{BL} denotes the theoretical expected frequency of persons with blue eyes and light hair, we have $E'_{BL} = nP(B \text{ and } L) = nP(B)P(L)$. Here as before, n is the total number of observations. However, the hypothesis of independence does not specify numerical values for $P(B)$ and $P(L)$. To compute a numerical value for E'_{BL} that can be used in a chi-square test, we estimate $P(B)$ and $P(L)$ from the available information as $\#(B)/n$ and $\#(L)/n$ respectively. We then estimate E'_{BL} as

$$E_{BL} = n \frac{\#(B)}{n} \frac{\#(L)}{n} = \frac{\#(B)\#(L)}{n}.$$

In our example, $E_{BL} = \dfrac{30 \times 40}{100} = 12.$

More generally, expected frequencies are obtained by multiplying the two marginal totals corresponding to a given cell in the two-way table and dividing the resulting product by n:

11.2
$$E = \frac{\text{(total for column)} \times \text{(total for row)}}{n}$$

Table 11.3 lists expected frequencies computed in this way in the lower right-hand corner of each cell. The number in the upper left-hand corner is the observed frequency. (The data in Table 11.1 are fictitious, chosen specifically to give round numbers for our marginal totals, thereby making the required computations much simpler. Real data are rarely as accomodating.)

TABLE 11.3

		eye color			
		brown	blue	grey	totals

hair color	light	13
		    20
			18
			    12
				9
				    8
					40

	dark	37
		    30
			12
			    18
				11
				    12
					60

	totals	50	30	20	100

Now we have the necessary information to compute x^2. But before we do so, we want to figure out the appropriate number of degrees of freedom. Recall that in the test of Chapter 10 the number of degrees of freedom was one less than the number of categories. For m categories, there were $m - 1$ degrees of freedom. Essentially, the reason was that we had to compute only $m - 1$ of the expected frequencies according to formula. The last, the mth expected frequency, could then be obtained by subtraction. But this time the situation is more complicated.

We notice that in Table 11.3 the expected frequencies add up to the same marginal totals as the observed frequencies. For example in the first row,

sum of expected frequencies $= 20 + 12 + 8 = 40$
$= $ sum of observed frequencies.

Thus after we have computed two of the expected frequencies, for example, 20 and 12, according to Formula 11.2, we can find the remaining expected frequency in the first row as the difference $40 - (20 + 12) = 8$ without referring to Formula 11.2. A similar situation holds for columns. Indeed in the present example all three expected frequencies in the second row can be obtained by subtraction without referring to Formula 11.2. For example, in the first column, $30 = 50 - 20$. We needed Formula 11.2 only twice, namely for the computation of the two theoretical frequencies 20 and 12 in the upper left and center cells. Consequently in the present example we have only two degrees of freedom.

The computation of x^2 itself is straightforward:

eye color	hair color	O	E	$O - E$	$(O - E)^2/E$
Br	L	13	20	−7	2.450
Bl	L	18	12	6	3.000
G	L	9	8	1	.125
Br	D	37	30	7	1.633
Bl	D	12	18	−6	2.000
G	D	11	12	−1	.083
totals		100	100	0	$9.291 = \chi^2$

From Table D we find that the computed value 9.291 of χ^2 with 2 degrees of freedom exceeds the .01 critical value of 9.21. We therefore decide to reject the hypothesis of independence of eye and hair color.

But what does this decision imply? Our hypothesis stated that there is no relationship between a person's hair color and eye color. Rejection of this hypothesis by the chi-square test suggests that a relationship in fact exists. How does this relationship manifest itself? A look at the differences $O - E$ suggests the following: Blue eyes and light hair and brown eyes and dark hair go together more frequently than predicted by independence; blue eyes and dark hair and brown eyes and light hair go together less frequently than is to be expected under independence.

71 Before we look at another example, we shall formulate our problem and its solution in general terms. Our sample consists of n items that we have selected at random from some underlying population. For each item we measure two characteristics. In the general case, we shall talk of characteristic A and characteristic B. Characteristic A has c categories $A_1, \ldots, A_i, \ldots, A_c$; characteristic B has r categories $B_1, \ldots, B_j, \ldots, B_r$.

The hypothesis that we want to test states that the two characteristics A and B are unrelated. Knowledge that a given item belongs to a certain A-category throws no light on the B-category to which the item belongs. The reverse is also true. In practice we often hope that this hypothesis is incorrect, and that knowledge of the A-category to which an item belongs gives some indication of the B-category to which it belongs.

A concise notation is helpful. For $i = 1, \ldots, c$ and $j = 1, \ldots, r$, we let O_{ji} stand for the number of observations that belong to the ith A-category and at the same time belong also to the jth B-category. The total number of observations in the ith A-category is denoted by S_i and the total number of observations in the jth B-category is denoted by T_j. We then have Table 11.4.

The expected frequency corresponding to O_{ji} is denoted by E_{ji}. Our earlier argument shows that $E_{ji} = S_i T_j/n$, the product of the two marginal totals divided by the total number of observations. Earlier considerations also show that the resulting χ^2 has $(r - 1)(c - 1)$ degrees of freedom, since the expected fre-

TABLE
11.4

$r \times c$ **Contingency Table**

	A_1	$\cdots$	A_i	$\cdots$	A_c	
B_1	O_{11}	$\cdots$	O_{1i}	$\cdots$	O_{1c}	T_1
$\vdots$	$\vdots$		$\vdots$		$\vdots$	$\vdots$
B_j	O_{j1}	$\cdots$	O_{ji}	$\cdots$	O_{jc}	T_j
$\vdots$	$\vdots$		$\vdots$		$\vdots$	$\vdots$
B_r	O_{r1}	$\cdots$	O_{ri}	$\cdots$	O_{rc}	T_r
	S_1	$\cdots$	S_i	$\cdots$	S_c	n

quencies in the last row and last column can be obtained by subtraction.

A Test of Homogeneity

72   As a further example, assume that classification A tells us whether a person has received no flu shots, exactly one flu shot, or more than one flu shot, while classification B tells us whether during a flu epidemic the person has or has not contracted flu. The hope, of course, is that flu shots reduce a person's susceptibility to flu and that multiple shots are more effective than a single shot. But is this hope really justified? We set up the null hypothesis that whether or not a person contracts flu is unrelated to how many flu shots he has received (if any). This hypothesis can be tested by means of the chi-square test previously discussed with $c = 3$ and $r = 2$.

The experimental setup in the flu experiment is likely to be different from the one involving eye and hair color. In that experiment, test subjects are presumably selected at random from a population of test subjects and each selected test subject is then classified with respect to eye and hair color. In the flu experiment, however, it is highly important that test subjects from all three immunization classes be included in the experiment, preferably in approximately equal numbers. It is then simpler to go out and collect *three* random samples, one from each immunization class, and classify the test subjects in each of the three samples according to whether they do or do not contract flu during the epidemic. The resulting frequencies can be exhibited in a contingency table with three columns (corresponding to the three immunization classes) and two rows (corresponding to "flu" or "no flu"). Strictly speaking it is then inappropriate to speak of the corresponding chi-square test as a test of independence, since a person's immunization status is not determined by chance. What we have in this case is a test of *homogeneity*. Let p_0 be the probability that a person who has received zero flu shots contracts flu, with p_1 and p_2 defined in a similar manner. To say that immunization status and flu experience are unrelated implies that there is no difference among these three probabilities, that is, $p_0 = p_1 = p_2$. This is the formal statement of the hypothesis of homogeneity: The three relevant probabilities are all equal.

73   In general, it is extremely complicated to answer medical questions such as the one we have just raised. To point out one difficulty in our flu example, it is not at all simple to establish reliably whether a person has had an actual case of the flu or just a heavy cold. Also, a person may have had a type of flu for which the flu vaccine had not been prepared. Another complicating factor is that some inoculated persons, while not completely immune, develop milder cases of the flu than they would have if they had not been inoculated. A careful investigation of the effectiveness of flu shots should be concerned with details of this type.

   Let us then illustrate the use of the chi-square test as a test for homogeneity with another more straightforward numerical example. Suppose that we are interested in finding out if, at a certain college, the choice of a student's major field of interest is related to his class in college. We ask samples of 100 freshmen, sophomores, juniors, and seniors at the college for their major field of concentration with the results given in Table 11.5. This is a typical case of testing for homogeneity. Is the proportion of natural science (social science, humanities) majors the same for all four college classes or are there differences? We can use a chi-square test to find out.

TABLE 11.5

field of concentration	freshman	sophomore	junior	senior	totals
natural sciences	45	36	26	28	135
social sciences	25	33	41	38	137
humanities	30	31	33	34	128
totals	100	100	100	100	400

class

   Expected frequencies are computed in the usual manner. For example, under the hypothesis that no differences exist, the expected number of freshmen who are natural science majors is $100 \times \frac{135}{400} = 33.75$. (While observed frequencies are of necessity integral numbers, it is bad practice to round off expected frequencies to the nearest integer.) Since all four samples are of the same size, the expected frequencies of sophomores, juniors, and seniors who are natural science majors are also 33.75. Correspondingly, we find the expected frequency of social science majors in any one year to be $100 \times \frac{137}{400} = 34.25$ and the expected number of humanity majors, $100 \times \frac{128}{400} = 32$. Thus

$$\chi^2 = \frac{(45 - 33.75)^2 + (36 - 33.75)^2 + (26 - 33.75)^2 + (28 - 33.75)^2}{33.75}$$

$$+ \frac{(25 - 34.25)^2 + (33 - 34.25)^2 + (41 - 34.25)^2 + (38 - 34.25)^2}{34.25}$$

$$+ \frac{(30 - 32)^2 + (31 - 32)^2 + (33 - 32)^2 + (34 - 32)^2}{32}$$

$$= 12.3$$

With $(r - 1)(c - 1) = (3 - 1)(4 - 1) = 6$ degrees of freedom, the result is not quite significant at the .05 level. If we strictly adhere to a 5% significance level, we have no reason to reject the hypothesis of homogeneity. However if we adopt a purely exploratory point of view, the data certainly put us on guard that the hypothesis may not be strictly correct. A comparison of observed and expected frequencies suggests that the proportion of natural science majors may be decreasing and the proportion of social science majors may be increasing as students progress from freshmen to seniors. The proportion of humanities majors seems to be rather stable (see Problem 6).

74 In carrying out this test of homogeneity we used Formula 11.2 to compute the expected frequencies. Since this formula was derived using the hypothesis of independence, and since here we have the hypothesis of homogeneity instead, we must check to see that the formula is still valid. We will do this for our specific example, but our reasoning can be applied to any similar test for homogeneity.

Let the subscripts 1, 2, 3, and 4 refer to class in college and let the subscripts N, S, and H refer to natural science, social science, and humanities respectively. Using this notation, we can define the twelve probabilities $p_{N1}, p_{N2}, \ldots, p_{H4}$, where, for example, p_{H2} is the probability that a sophomore student majors in the humanities. The precise hypothesis of homogeneity then states that

$$p_{N1} = p_{N2} = p_{N3} = p_{N4},$$
$$p_{S1} = p_{S2} = p_{S3} = p_{S4},$$
$$p_{H1} = p_{H2} = p_{H3} = p_{H4}.$$

Let p_N denote the common probability that a student is a natural science major, with p_S and p_H defined correspondingly. If n_1, n_2, n_3, and n_4 are the four sample sizes (in our example, $n_1 = n_2 = n_3 = n_4 = 100$), the expected frequency of, say, sophomore humanities majors is $E'_{H2} = n_2 p_H$. However, as with the hypothesis of independence, the hypothesis of homogeneity does not specify a value for p_H. We therefore estimate p_H from the given information as $\#(H)/n$, where $\#(H)$ is the total number of humanities majors in all four samples and $n = n_1 + n_2 + n_3 + n_4$ is the total number of observations. E'_{H2} is then estimated as

$$E_{H2} = n_2 \frac{\#(H)}{n} = \frac{(\text{column total}) \times (\text{row total})}{n}$$

as before. A similar argument can be used for each of the other expected frequencies, so we see that Formula 11.2 is also valid for tests of homogeneity. Note also that we have the same number of degrees of freedom for the test of homogeneity as for the test of independence, namely $(r-1)(c-1)$, since the marginal totals impose the same restrictions on the expected frequencies in both cases.

PROBLEMS *In problems where the hypothesis that is being tested is rejected, the student should indicate what alternative is suggested by the data.*

1 A final examination was taken by 35 male students and 31 female students. The instructor noted whether a student was sitting in the front rows or the back rows with the following results:

	front rows	*back rows*
male students	12	23
female students	24	7

Assume that the 66 students represent a random sample from the population of students taking the type of course involved. Set up an appropriate hypothesis and test it.

2 At a university, students from two different colleges take the same calculus examination. Here are the grades:

	grades				
	A	B	C	D	F
college 1	6	13	43	16	22
college 2	18	26	41	6	9

Set up an appropriate hypothesis and test it.

3 In an experiment, 50 smokers and 50 nonsmokers are asked whether they believe that heavy smoking may lead to lung cancer and other serious diseases. Their answers are tabulated as follows:

	do believe	*do not believe*
smokers	11	39
nonsmokers	28	22

Set up an appropriate hypothesis and test it.

4 At the University of California, 643 women students were classified with respect to looks and year of college attendance:

	homely	*plain*	*good-looking*	*beautiful*
junior	17	68	108	25
senior	27	100	84	33
graduate	39	82	44	16

Source: *Human Biology*, 10 (1938), 65–76.

Set up an appropriate hypothesis and test it.

5 In a pre-election poll studying the influence of age on voter preference for two presidential candidates, the following results were obtained:

	prefer candidate A	prefer candidate B	undecided
20–29	67	117	16
30–49	109	74	17
over 49	118	64	18

Set up an appropriate hypothesis and test it.

6 The data in Table 11.5 can be contracted as follows:

	freshman	sophomore	junior	senior
other than humanities	70	69	67	66
humanities	30	31	33	34

Test the hypothesis that the choice of a humanities major is unrelated to the class of the student.

12

One-Sample
Methods

Continuous Measurements

75   The statistical methods we have discussed so far assume that an observation can be classified as belonging to one of a relatively small number of categories. The kind of questions that customarily arise in connection with categorical data are rather simple. What proportion of the items in the population as a whole belong in each category? Is it correct to assume that the items in the population are evenly divided among the m categories? If there are two criteria for classifying a given item, are the two criteria related, and if so, how? Our basic information for answering such questions consists of counts of items in the various categories, and the appropriate statistical methods are often classified as *enumeration statistics*.

As long as the subject matter under investigation is of a purely categorical or qualitative nature, we are limited very much to the methods of statistical analysis discussed so far. But often observations provide quantitative information. In such cases more complicated questions can be asked. We shall raise such questions in the remaining chapters and try to find some answers. Before we do so, however, some ground work is required.

In the case of categorical data all problems revolve around a finite number of probabilities, namely, the probabilities associated with the various categories. But a finite number of probabilities is not sufficient to describe the distribution of measurements such as the number of miles per gallon of gasoline a person gets from his car, or the weight of a person, or the number of hours of viewing that a television tube provides before burning out.

For example, if you ask a person's weight, you are very likely to get an answer like 143 or 160 or 275 pounds. In a sense these answers can be considered as categorical data, similar to the categories 2, 3, ..., 12 for the sum of points on two dice. But there is an important difference between the two cases in addition to the fact that the number of weight categories is much larger than the number of dice results. In the case of weights, the results are rounded off. In the case of dice, the results are exact. Thus if a person says that his weight is 143, he presumably implies that his true weight is somewhere between $142\frac{1}{2}$ and $143\frac{1}{2}$ pounds. People have a tendency to round off their weight to the nearest pound or possibly even to the nearest 5 pounds. In the case of the person who says that his weight is 275, the rounding off may even have been to the nearest 25 pounds.

When two dice are rolled, the resulting sum of points can only be one of a finite number of isolated values, namely, 2, 3, ..., 12. The weight of a person, on the other hand, can be any number in a certain range. In the dice example it makes sense to talk of the probability that the sum of points is exactly 10. In the case of the weight of a person, it is trivial to talk of the probability that a person's weight is exactly 150 pounds, meaning 150.00 ... pounds. This probability is zero. A probability that is not trivial is the probability that a person's weight is between 145 and 152 pounds.

The number of points on two dice is a *discrete* quantity. The weight of a person is a *continuous* quantity. For discrete quantities it is sufficient to know the probabilities associated with the individual values that can occur. In the continuous case we must be able to determine the probabilities associated with all possible intervals.

This raises new mathematical problems that can be solved rigorously only with the help of integral calculus. However, in this book we have neither the time nor the inclination to look in detail at the purely mathematical side of the problem. In Section 34 when discussing the binomial distribution we represented probabilities as areas under a histogram and then in turn approximated the histogram by a normal distribution. At that time we pointed out that the significance of normal distributions as a tool for computing probabilities goes much beyond problems involving binomial trials. Experience shows that the distribution of many kinds of measurements, like the weights of persons, gasoline mileages, and test scores, can be very adequately described by suitably chosen normal distributions.

There was a time when it was thought that measurements that could not be described by a normal distribution were abnormal and subject to suspicion. Nowadays we realize that this is not the case. While certain phenomena can adequately be described using a suitable normal distribution, others cannot. For example, when we study yearly family income, we find that the distribution of income can be described by a curve like the one in Figure 12.1. Most families have incomes that fall in a relatively limited range, but a small proportion of families have incomes spreading far beyond

FIGURE
12.1

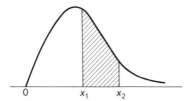

this range. The interpretation of a curve like the one in Figure 12.1 is, however, no different from that of a normal distribution. The area under the curve between two ordinates x_1 and x_2 equals the proportion of families with yearly incomes between x_1 and x_2 dollars.*

In general, then, continuous measurements are described probabilistically in terms of a suitable distribution function. Mathematically a distribution function is determined by its equation $y = f(x)$. For example, the normal distribution with mean μ and standard deviation σ has the forbidding looking equation

$$f(x) = \frac{1}{\sigma\sqrt{2\pi}} e^{-\frac{1}{2}\left(\frac{x-\mu}{\sigma}\right)^2}.$$

(Fortunately we do not have to remember such things.)

Nonparametric and Normal Theory Methods

76 In practice it rarely happens that the equation of the appropriate distribution function is completely known. Often theory or past experience or both suggest a distribution function of a certain type, for example, the normal type, without specifying the exact values of parameters like μ and σ that enter into the equation. A great deal of present day statistics is concerned with problems arising from the estimation of and tests of hypotheses involving parameters of one or several normal distributions. In the hands of a skilled statistician, these normal theory methods are immensely valuable. In the hands of a beginner, they may produce rather misleading results. We may compare normal theory methods to a rather expensive camera with all kinds of extra equipment. In the hands of an experienced photographer, the camera gives results that are usually worth the price. But beginners only get confused and discouraged and in most cases would be much better off with a simple box camera.

In Chapters 17–21 we shall study normal theory methods. But now we first take up some "box-type" procedures. These procedures are conceptually much simpler than normal theory

---

*In mathematical treatments it is customary to refer to curves like the one pictured in Figure 12.1 as *density* functions. We shall use the more general term *distribution* function instead, even though the term distribution function often implies a cumulative distribution function.

methods. They also afford the user much more latitude in his assumptions. The word *robust* has been used to describe certain of their properties. They are most commonly referred to as *nonparametric* methods, even though they sometimes do concern themselves with parameters, as we shall see shortly.

There is always the danger of reading too much into a comparison like our comparison between statistics and photography. Skilled photographers hardly ever use box cameras. Skilled statisticians more and more frequently use nonparametric methods. Both nonparametric and normal theory methods have their rightful place in a statistician's tool kit. On the whole, beginning statisticians are less likely to go wrong using nonparametric methods than using normal theory methods.

We mentioned that nonparametric methods afford the user greater latitude in his assumptions. This statement requires elaboration. As early as Chapter 1 when discussing the taxi problem, we saw that statistical methods are not universal in their application. Our estimate of the number T of taxis could be expected to give useful information only if taxis were numbered from 1 to T (aside from the assumption of randomness of our sample, an assumption that is implicit for all methods discussed in this book). If the assumption that taxis are numbered consecutively from 1 to T is inappropriate, our estimate of T may be greatly in error.

The normal theory methods to be discussed in Chapters 17–21 assume that the phenomena under investigation can adequately be described by means of normal distributions. If this assumption is satisfied, normal theory methods are the best possible ones we can use. No other methods provide more accurate estimates of unknown parameters or smaller error probabilities in tests of hypotheses.

However, not all phenomena producing measurements on a continuous scale follow normal distributions. Clearly, any asymmetry implies nonnormality. Even symmetry is no assurance of normality, for there exist phenomena of a symmetric nature that cannot adequately be described by means of normal distributions. The nonparametric methods that we shall discuss in this and the next four chapters are appropriate whenever observations may reasonably be assumed to have come from some continuous population. Since the assumption of normality implies continuous measurements, nonparametric methods are applicable whenever normal theory methods are. The reverse is, of course, not true. There are situations in which the use of nonparametric methods of analysis is appropriate while that of normal theory methods may produce rather misleading conclusions.

Estimating the Median of a Population

77   *The Median as a Measure of Centrality*   We mentioned earlier that in order to describe a population completely, we have to know its distribution function $f(x)$. Since in general $f(x)$ is unknown, we usually have to be satisfied with considerably less information. Thus, a few numbers that characterize important aspects of the distribution are often all we can hope to investigate. Some kind of central value would seem to be of greatest interest. In the case of a symmetric distribution we clearly should like to know where the center of symmetry is located. But what value can we use to characterize the center of an asymmetric distribution like that of yearly family incomes? What shall we call a typical income? While there is no unique answer to this question, official reports often quote the *median* income. The median income divides all incomes in two equal parts. There are as many families with incomes higher than the median income as there are families with incomes lower than the median income. More generally, the median η of a distribution $f(x)$ divides the total area under the distribution into two equal parts, as in Figure 12.2. Of course for symmetric distributions, the center of symmetry and the median coincide.

FIGURE 12.2

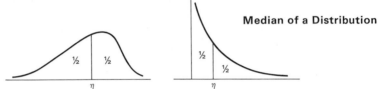

Median of a Distribution

We have the following problem. We are interested in estimating the median η of a certain population, but we do not know the precise distribution function $f(x)$. On the other hand, we are able to observe a random sample of n observations which we denote by $x_1, \ldots, x_n$. We want to use these observations to estimate η, first by a point estimate, and then by means of a confidence interval.

Specifically, we consider the following situation. We are interested in studying the earnings of students at a large university during the school year. Because of the large number of students, it is impractical to ask each one about his earnings. But by the methods discussed earlier, we can select a random sample of, say, 100 students and find out what their earnings are. In this case, $n = 100$, x_1 denotes the earnings of the first student in our sample, x_2, those of the second student, and so on.

A rather obvious point estimate of the population median is the sample median M, the median of the set of observations $x_1, \ldots, x_n$. In the student earnings example we would arrange the various earnings from the lowest to the highest and then find the number halfway between the 50th and 51st observation in this rearranged series.

78 *A Confidence Interval for the Median*   Our point estimate of η has the usual disadvantages of point estimates in general. Without supplementary information we have no idea of its accuracy. And in the case of the median, supplementary information is difficult to come by. On the other hand, it is very easy to find a confidence interval for η. The length of a confidence interval provides a built-in indicator of its accuracy. A short interval provides precise information; a long interval provides only vague information.

By far the simplest interval is obtained by using the smallest and the largest observations in the sample as lower and upper limits, that is,

$$L_1 \leq \eta \leq U_1,$$

where L_1 is the smallest sample observation and U_1, the largest. What we have to know is how confident we can be in this statement. In other words, what is the confidence coefficient γ associated with this interval? The answer to this question is quite simple. Our statement is correct unless our n observations are either all greater than the population median η or all smaller than η. Since the median η divides the population in two equal parts, the probability that a single observation is greater than η is $\frac{1}{2}$. It follows that the probability that all n observations are greater than η is $(\frac{1}{2})^n$. Similarly, this is also the probability that all observations are smaller than η. Thus

$$\gamma = 1 - 2(\tfrac{1}{2})^n = 1 - \frac{1}{2^{n-1}}.$$

Table 12.3 illustrates this formula.

Confidence Coefficients of Confidence Intervals for Median

n	2	3	4	5	6	7	8
γ	.500	.750	.875	.938	.969	.984	.992

We can be at least 99% confident that the population median lies between the smallest and the largest observations in a sample containing 8 observations. And if we are satisfied with a 97% confidence coefficient, we need only 6 observations.

By taking samples containing more than 8 observations we can be even more confident in the correctness of our statements. But there is rarely any practical need for having a confidence interval with confidence coefficient greater than, say, .99 or .995. And, of course, we are buying this high assurance of the correctness of our statement at an expense. It is correct that as the number of observations increases, our method produces confidence intervals with higher and higher confidence coefficients. But at the same time, the length of the confidence interval tends to increase together with the number of observations. The smallest and the largest observations in a sample of size 8 are likely to be considerably further apart than the smallest and largest observations in a sample of size 4. While a higher confidence coefficient means a more reliable statement, a longer interval means a less practically useful statement. We have noticed similar situations before in connection with confidence intervals for the parameter p of a binomial distribution and in the inverse relationship between probabilities of type 1 and type 2 errors in testing statistical hypotheses. In determining a confidence interval for any parameter, we have to strike a balance between the confidence coefficient we should like to achieve and the expected length of the resulting confidence interval. What, then, can we do in this case?

79 The confidence interval bounded by the smallest and largest sample observations is as simple an interval as we can find. But almost as simple is an interval bounded by the second smallest and second largest observations in the sample. More generally, let

$$L_d = d\text{th smallest sample observation,}$$
$$U_d = d\text{th largest sample observation.}$$

We can then consider confidence intervals of the type

$$L_d \leq \eta \leq U_d.$$

The computation of the corresponding confidence coefficient γ is only slightly more complicated than in the case $d = 1$, which we considered earlier (see Problem 11). However, we shall simply use Table E which, for various values of d, gives the confidence coefficient γ associated with the confidence interval $L_d \leq \eta \leq U_d$.

NOTE 12.4 We have mentioned before that many statisticians prefer routine confidence coefficients such as .90, .95, or .99. In general the methods of this and the next two chapters do not permit us to find intervals corresponding exactly to one of these confidence coefficients. For this reason Table E (as well as Tables F and G) lists d-values such that the corresponding confidence coefficients bracket routine levels .99, .95, and .90. The statistician can then select the d-value that best satisfies his intentions.

EXAMPLE 12.5a Nine randomly selected students obtained the following scores on a certain examination:

77, 88, 85, 74, 75, 62, 80, 70, 83.

Find a point estimate and confidence interval for the median score for this examination.*

The sample median is 77. This is our point estimate. According to Table E a confidence interval with confidence coefficient .961 is bounded by the second smallest and the second largest of the observations. Thus we find the confidence interval

$$70 \leq \eta \leq 85.$$

---

*In practice we would either find a point estimate *or* a confidence interval. Here we are interested in illustrating both procedures.

As grades go, this is a rather wide interval, indicating that a sample of size 9 does not provide very precise information.

80   *Symmetric Populations*   The confidence intervals for medians that we have just discussed are typical of nonparametric methods, with respect to both conceptual and computational simplicity. More importantly, however, the results are valid without any assumptions about the population distribution $f(x)$. Of course, if special knowledge about $f(x)$ is available, more accurate methods are often possible. Thus it may be reasonable to assume that $f(x)$ is symmetric. As we have already observed, the median η then coincides with the center of symmetry. And in finding an estimate for η, it is often advantageous to make use of the symmetry of $f(x)$.

When $f(x)$ is symmetric, the sample observations $x_1, \ldots, x_n$ will be more or less symmetrically positioned around η. The same will also be true of the $n(n-1)/2$ averages of two observations,

$$\frac{x_1 + x_2}{2}, \frac{x_1 + x_3}{2}, \ldots, \frac{x_{n-1} + x_n}{2}.$$

Let us add to this list of averages the original observations $x_1, \ldots, x_n$. Since we can write $x_i = (x_i + x_i)/2$, we find it convenient to use the notation

$$u_{ij} = \frac{x_i + x_j}{2}, \ 1 \leq i \leq j \leq n.$$

The median of the $n(n-1)/2 + n = n(n+1)/2$ u_{ij}'s is an estimate of the center of symmetry η that is usually better than the sample median. A corresponding confidence interval for η can be found in much the same manner as in the preceding procedure. As the lower confidence limit we use the dth smallest u_{ij}; as the upper limit, the dth largest u_{ij}. This time the corresponding confidence coefficient is read from Table F.

EXAMPLE
12.5b

Examination scores are usually rather symmetrically distributed. We shall therefore apply the new procedures to the data of Example 12.5a. Let us first find a confidence interval. According to Table F, a confidence interval with confidence coefficient .961 is bounded by the 6th smallest and 6th largest of all u_{ij}'s. It is helpful to rearrange the observations according to size:

$$62, 70, 74, 75, 77, 80, 83, 85, 88.$$

Since 62 averaged with each of 62, 70, 74, 75, and 77 produces an average that is smaller than 70 but $\frac{1}{2}(62 + 80) = 71$, the 6th smallest average is the number 70 itself. In a similar fashion we find that the 6th largest average is 84. Thus our confidence interval is $70 \leq \eta \leq 84$, slightly shorter than the earlier interval. But it should be remembered that we are making use of an additional assumption, namely the symmetry of the distribution of grades. If the assumption of symmetry is unjustified, we have no right to use the second confidence interval.

The search for the point estimate of η is more time consuming. There are $9 \times 10/2 = 45$ u_{ij}'s. Therefore the point estimate is given by the 23rd smallest (or largest) u_{ij}. We may proceed as follows. The 5 smallest observations (62, 70, 74, 75, 77) produce 15 averages, each of which is clearly smaller than or equal to 77. In addition, 62 averaged with each of the remaining 4 observations (80, 83, 85, 88) produces a u_{ij} that is smaller than or equal to 77. The same is true of 70 averaged with 80 and 83, and 74 averaged with 80. By now we have accounted for $15 + 4 + 2 + 1 = 22$ u_{ij}'s. The 23rd u_{ij} then equals $(75 + 80)/2 = 77.5$, and this is our point estimate of η.

81   A systematic ordering of the u_{ij}'s can be achieved graphically as follows. On a sheet of graph paper we plot the data points along the 45° line starting at the lower left-hand corner. (A simple method for doing so is to plot the data points along the vertical axis on the left and project the resulting points horizontally onto the 45° line.) Below the 45° line we mark all intersections of parallels to the horizontal and vertical axes through the n data points. These intersections together with the data points on the 45° line represent the u_{ij}'s. The intersections are ordered according to increasing (decreasing) size of u_{ij} by sliding a line perpendicular to the 45° line from below (above) and counting intersections including data points. A draftsman's triangle made of clear plastic is very helpful in performing the counting operation.

EXAMPLE 12.5c   The necessary steps for the graphic solution of Example 12.5b are indicated in Figure 12.5d. Sliding the perpendicular to the 45° line from below, we find that the 6th intersection corresponds to the data point 70. Thus the lower confidence limit is 70. Sliding the

FIGURE 12.5d

Estimation of Median

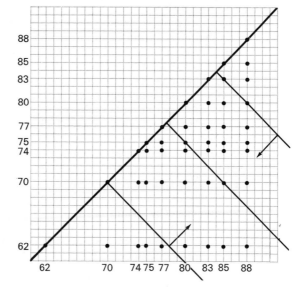

perpendicular from above, the 6th intersection is formed by the horizontal through 80 and the vertical through 88. Thus $u_{ij} = (80 + 88)/2 = 84$. This is the upper confidence limit. We obtain the same result if we use the intersection of the horizontal through 83 and the vertical through 85. The point estimate of η corresponds to the 23rd intersection from above or below. Again we have two choices, $(70 + 85)/2 = 77.5 = (75 + 80)/2$.

Tests of Hypotheses about Medians

82   In some situations, rather than finding a point or interval estimate for the population median, we may want to test a hypothesis. For example, we may have heard a claim that the median annual family income in a certain community is $10000 and are curious about its accuracy. We need a test of the hypothesis

$$\eta = \eta_0 \ (= 10000, \text{ in our example}).$$

In connection with the binomial distribution, we saw that a hypothetical success probability p_0 can be tested at significance level α by finding out whether or not a corresponding confidence interval with confidence coefficient $\gamma = 1 - \alpha$ contains the hypothetical parameter value. Similarly, we can test the hypothesis $\eta = \eta_0$ by finding out whether or not a confidence interval for η contains the value η_0. We have discussed two different methods for finding a confidence interval for the population median η. The method discussed in Section 79 is completely general, while the method discussed in Section 80 assumes that the population is symmetric about the population median. Depending on whether or not we are willing to assume symmetry, we then have appropriate methods for testing the hypothesis $\eta = \eta_0$.

As we have observed before, a confidence interval contains *all* η-values that are acceptable. On the other hand, there are times when we are exclusively interested in the acceptability or non-acceptability of one and only one value η_0. Can we test the hypothesis $\eta = \eta_0$ without finding a confidence interval first? Indeed we can. We shall see presently how the two confidence procedures can easily be converted into test procedures.

Direct test procedures are particularly useful when we are interested in one-sided alternatives. The test that rejects the hypothesis $\eta = 10000$ when 10000 is not in the confidence interval is a test against the two-sided alternative $\eta \neq 10000$. But we may actually doubt that the median income is as high as $10000. We are then really interested in a test of the hypothesis $\eta = 10000$ against the alternative $\eta < 10000$. While it is possible to modify confidence intervals to provide tests against one-sided alternatives (by using only one of the two bounds—in our example, we would use only the upper bound—and changing the confidence coefficient correspondingly), a direct test procedure is often more convenient.

83 *The Sign Test* Since our first confidence interval for η is bounded by the dth smallest and the dth largest observations, a hypothetical value η_0 must be either smaller than the dth smallest observation or larger than the dth largest observation in order *not* to be acceptable. In the first case, fewer than d observations are smaller than η_0. In the second case, fewer than d observations are larger than η_0. The hypothesis that $\eta = \eta_0$ can then be tested in the following way without first finding a confidence interval for η. Let S_- equal the number of observations that are smaller than η_0 and S_+, the number of observations that are greater than η_0; set S equal to the smaller of S_- and S_+. The hypothesis $\eta = \eta_0$ is rejected if S is *smaller* than d. This test is known as the *sign test*, since it simply counts how many of the differences $x_i - \eta_0$ are negative (whenever x_i is smaller than η_0) and how many are positive (whenever x_i is greater than η_0).

The procedure that we have just described provides a two-sided test, that is, a test against the alternative $\eta \neq \eta_0$. The significance level of this two-sided test is $1 - \gamma$. A one-sided test is performed by using as test statistic either S_- or S_+, whichever can be expected to be smaller under the alternative hypothesis. Thus under the alternative $\eta > \eta_0$, we expect few observations to be smaller than η_0, and therefore use S_- as the test statistic. On the other hand, under the alternative $\eta < \eta_0$, we expect few observations to be greater than η_0, and therefore use S_+ as our test statistic. The significance level of a one-sided test is $(1 - \gamma)/2$, one half the significance level of the two-sided test. Table E, in addition to listing confidence coefficients γ, lists significance levels $\alpha'' = 1 - \gamma$ for two-sided tests and $\alpha' = (1 - \gamma)/2$ for one-sided tests. Note that, due to rounding, we need not necessarily have $\alpha'' = 2\alpha'$ exactly. For example, each of the one-sided significance probabilities .0496, .0502, and .0504 would be listed as $\alpha' = .050$. However, the corresponding two-sided significance probabilities would be listed as $\alpha'' = .099, .100$, and $.101$, respectively.

The sign test procedure assumes that none of the observations equals η_0. Any observations that are equal to η_0 should be eliminated from the sample and the sample size should be adjusted before carrying out the sign test.

EXAMPLE 12.6 In order to test the hypothesis that the median yearly income of families in a certain community is $10000 (or more) against the alternative that it is less than $10000, we have obtained the incomes of a random sample of 20 families:

9300	50000	15500	7000	12700
10000	9800	21000	6900	9100
8200	7600	9800	9300	35000
6500	5900	9200	8500	8000

Suppose we want a test with significance level as close to .05 as we can get. One observation equals the hypothetical value 10000. After eliminating this observation we have 19 observations left.

According to Table E, for $n = 19$, a one-sided test with $d = 6$ has significance level .032. In our case the appropriate statistic is $S_+ = \#(\text{observations} > 10000) = 5$. Since $S_+ < 6$, we reject the hypothesis at significance level .032.

Table E provides critical values for the sign test corresponding to routine significance levels for samples with $n \leq 50$. Actually, unless n is quite small (say smaller than 10 or 12), we can use the normal distribution to compute approximate probabilities. Suppose we use S_- as our test statistic. We can then call an observation x_i that is smaller than the hypothetical median η_0 a success and an observation that is greater than η_0 a failure. Thus, S_- is the number of successes in n independent trials, and exact probabilities involving S_- (and similarly, S_+) are obtained from the appropriate binomial distribution. In particular, when the null hypothesis is true, that is, when the true population median is η_0, the success probability is $\frac{1}{2}$. The mean and standard deviation to be used in the normal approximation are then

$$\mu = np = \tfrac{1}{2}n \text{ and } \sigma = \sqrt{npq} = \tfrac{1}{2}\sqrt{n}.$$

EXAMPLE 12.7 We want to test the hypothesis $\eta = \eta_0$ against the one-sided alternative $\eta > \eta_0$. Suppose that in a sample of size 36 we find $S_- = \#(\text{observations} < \eta_0) = 12$. The descriptive level of this result using the normal approximation is

$$P(S_- \leq 12) = \phi\left(\frac{12 + \tfrac{1}{2} - \tfrac{1}{2} \times 36}{\tfrac{1}{2}\sqrt{36}}\right) = \phi(-1.83) = .034.$$

According to Table E, the exact descriptive level is $P(S_- \leq 12) = P(S_- < 13) = P(S_- < d) = .033$. Of course for tests against the two-sided alternative $\eta \neq \eta_0$, these probabilities should be doubled.

84 *Ranks* Before we can describe the test procedure corresponding to the confidence interval for η in the symmetric case, we have to define an important new term. Most nonparametric tests are based on *ranks* rather than actual observations. The rank of a given observation is obtained by arranging all observations according to size and giving the smallest observation the rank 1, the next smallest, the rank 2, and so on. The largest of n observations receives the rank n.

This procedure assumes that no two observations are equal, or *tied*, as we shall say from now on. If *ties* do occur among the observations, we use average ranks or *midranks*. Thus if we have the 5 observations

$$6, 8, 8, 11, 15,$$

we would assign the ranks

$$1, 2.5, 2.5, 4, 5,$$

since the two equal observations share the second and third positions when all observations are arranged according to size. Again, for the 5 observations

$$8, 8, 15, 11, 8,$$

we use ranks

$$2, 2, 5, 4, 2,$$

since the three equal observations share the first, second, and third positions when the observations are arranged according to size.

85 *The Wilcoxon Signed Rank Test* If the sample observations $x_1, \ldots, x_n$ may be assumed to have come from a symmetric distribution with center of symmetry η, a confidence interval for η is bounded by the dth smallest and dth largest of the $n(n + 1)/2$ averages $u_{ij} = (x_i + x_j)/2$, $1 \leq i \leq j \leq n$, where d is read from Table F. As we have already pointed out, this confidence interval can be used to test the hypothesis $\eta = \eta_0$. Alternatively, we can use the following direct procedure. In analogy to the sign test, we find the two statistics T_- and T_+ defined below and set T equal to the smaller one of the two. The hypothesis $\eta = \eta_0$ is rejected if T is *smaller* than d. This is a two-sided test with significance level $1 - \gamma$ ($= \alpha''$ in Table F). For the one-sided alternative $\eta > \eta_0$ we always use the statistic T_-; for the alternative $\eta < \eta_0$, the statistic T_+. These one-sided tests have significance level $(1 - \gamma)/2$ ($= \alpha'$ in Table F).

Theoretically T_- and T_+ are defined in terms of the number of u_{ij}'s that are smaller than, equal to, or larger than the hypothetical median. However there is a more convenient way for computing T_- and T_+ that does not require finding the individual u_{ij}'s. Instead we find all differences $x_i - \eta_0$. The absolute values $|x_i - \eta_0|^*$ are then ranked from 1 to n. T_- equals the sum of the ranks corresponding to negative differences $x_i - \eta_0$; T_+, the sum of the ranks for positive differences $x_i - \eta_0$.

Any observations that equal η_0 are omitted from the sample, exactly as in the sign test. Of course, the sample size has to be reduced correspondingly. We then always have $T_- + T_+ = 1 + 2 + \cdots + n = n(n + 1)/2$ where n equals the number of non-zero differences $x_i - \eta_0$, so that only one of the two rank sums need be computed from the data. The other rank sum can be obtained by subtraction.

This test is known as *Wilcoxon's signed rank test*.

EXAMPLE 12.8 A machine is supposed to produce wire rods with diameter 1 mm. In order to check whether the machine is properly adjusted, 12

---

*The absolute value of a number (indicated by vertical bars) is its numerical value without regard to sign. Thus $|+5| = |-5| = 5$.

rods from the machine's production are selected and measured giving the following results: 1.017, 1.001, 1.008, .995, 1.006, 1.011, 1.009, 1.009, 1.003, .998, .990, 1.007. Would you say that the machine needs adjustment?

We set up a test of the hypothesis $\eta = 1.000$ against the alternative $\eta \neq 1.000$. If this hypothesis is rejected, machine adjustments are indicated. Industrial measurements of this type are generally symmetrically distributed. We therefore use the Wilcoxon signed rank test. The necessary computations follow:

x_i	$1000\|x_i - 1.000\|$*	rank
1.017	17	12
1.001	1	1
1.008	8	7
.995	5	4
1.006	6	5
1.011	11	11
1.009	9	8.5
1.009	9	8.5
1.003	3	3
.998	2	2
.990	10	10
1.007	7	6

Since there are only three negative differences $x_i - \eta_0$, we compute $T_- = 4 + 2 + 10 = 16$. Clearly $T = T_- = 16$ (since $T_+ = n(n + 1)/2 - T_- = 78 - 16 = 62 > 16$). According to Table F the result is not significant at the .05 level. The significance level is approximately .08. A check-up on the machine adjustment may be advisable.

As in the case of the sign test, approximate probabilities can be obtained with the help of the normal distribution. For the Wilcoxon signed rank test, $\mu = n(n + 1)/4$ and $\sigma = \sqrt{n(n + 1)(2n + 1)/24}$.

EXAMPLE 12.9 We have computed $T_+ = 78$ in a sample of $n = 24$ observations. What is the descriptive significance level associated with this result? We find $\mu = (24 \times 25)/4 = 150$ and $\sigma = \sqrt{(24 \times 25 \times 49)/24} = 35$ and therefore $P(T_+ \leq 78) = \phi\left(\dfrac{78 + \frac{1}{2} - 150}{35}\right) = \phi(-2.04) = .021$. This is the descriptive level for a one-sided test. For a two-sided test it is .042. Note that Table F only tells us that the two-sided descriptive level is between .011 and .049. More accurate information requires interpolation.

86 Occasionally we have analyzed (or shall analyze) the same set of data by two or more methods. This is being done merely for illustrative purposes. The student should not draw the conclusion

---

*We multiply by 1000 in order to avoid decimals. This does not change the ranking.

that in practical work it is appropriate to analyze a given set of observations by various methods and *then* select the results that appear most satisfactory under the circumstances (for example, select the shortest confidence interval). When planning an experiment, the statistician should decide what method of analysis is to be used on the data resulting from the experiment. Thus in the case of symmetric populations, when both the Wilcoxon signed rank test and the sign test are appropriate, the statistician would most likely decide on the former, since it specifically uses symmetry, unless he feels that the greater simplicity of the sign test outweighs any considerations of greater accuracy.

A different situation prevails in the initial stages of a new investigation. The investigator may not have any mathematical model in mind but is performing the experiment in order to gain new insights. He may then subject his data to many different kinds of analyses. Here the purpose is not to reach any final conclusions but to develop a feeling for the kind of phenomena with which he is dealing. In recent years statisticians have used the term *data analysis* to refer to this kind of exploratory investigation. Before a statistician can be successful at data analysis, he must be familiar with standard statistical procedures.

Nonparametric Procedures and Tied Observations

87   The earlier mentioned restriction of nonparametric procedures to continuous data requires some comments. Nonparametric procedures when applied to continuous data are *distribution-free;* when applied to discrete data, they are not distribution-free. By this we mean the following. The confidence coefficients and significance levels listed in Tables E and F (as well as in similar tables) are exact whatever the sampled population, as long as this population is continuous. For example, the confidence interval for the population median bounded by the second smallest and second largest in a sample of eight observations is .930 whether the sample comes from a population with a distribution like those in Figure 12.2 or from a normal distribution. The confidence statement is "free" of any distributional assumptions beyond the assumption of continuity. A corresponding statement holds for the significance level of a nonparametric test like the sign test or the Wilcoxon signed rank test. The same is not true when the underlying population is discrete. In that case, the exact value of a confidence coefficient or significance level depends on the usually unknown probabilities associated with the various population values. This does not mean that nonparametric methods are inapplicable when analyzing data from discrete populations, as is sometimes stated. As we shall see, even in the discrete case statements that are practically useful are possible. However before we explain the nature of these statements, it is important to know the reason for the different behavior of nonparametric procedures when applied to continuous and when applied to discrete populations.

Theoretically, when sampling from a continuous population, we can be certain that no ties occur among the observations. However, in practice no phenomenon produces measurements of a strictly continuous nature. For example, the weight of a person can be any number within a certain range, but due to natural limitations of the accuracy with which a person's weight can be determined, actual weights are stated as 140 or 143 or 143.2 or even 143.18 pounds. These are discrete values. With yearly family incomes, the most accurate statement made (and even the Bureau of Internal Revenue does not require such accuracy) can only be to the nearest cent. On the other hand, measurements like test scores are naturally discrete, but it is often more convenient to treat them as continuous data. Because of this underlying discreteness, ties occur with positive probability when sampling from these populations. The presence of occasional ties among the observations complicates the application of nonparametric procedures.

88   What modifications are required, then, when applying nonparametric methods to discrete data? Let us first consider confidence intervals. In particular, for the sake of concreteness, consider the confidence interval for the median η,

$$L_d \leq \eta \leq U_d.$$

We stated the confidence interval as a *closed* interval including the endpoints as acceptable parameter values. If the underlying population is strictly continuous, the confidence coefficient is not changed if we remove the endpoints from the interval of acceptable parameter values, that is if we consider the *open* interval

$$L_d < \eta < U_d.$$

The situation is different when the population is discrete. The true confidence coefficient associated with the open interval may then be smaller than that associated with the closed interval. More specifically, the open interval has a true confidence coefficient that is at most equal to the tabulated value γ; the closed interval has a confidence coefficient that is at least equal to the tabulated value γ. Corresponding statements hold for other nonparametric confidence intervals. From a practical point of view then, it is always safer to use a closed interval. This is the rule we shall follow. As a consequence, true confidence levels will at least equal quoted confidence levels.

89   As far as confidence intervals are concerned, we use exactly the same procedure whether or not ties are present among the observations. Only the probability statement, that is, the statement involving the confidence coefficient, has to be modified. The situation is rather different when it comes to tests of hypotheses. Here the test procedure itself usually requires modification. The Wilcoxon signed rank test is a good example to illustrate the point.

When there are no ties among the observations, or rather among the absolute differences $|x_i - \eta_0|$, we only need the simple ranks $1, 2, \ldots, n$ when computing the Wilcoxon signed rank test statistic T. The need for midranks arises only when ties are present. Thus the extension of the definition of ranks to include midranks is due to a desire to be able to apply nonparametric tests even when there are ties among the observations. The rule that observations that equal the hypothetical parameter value η_0 be omitted from the analysis in both the sign and Wilcoxon signed rank tests is another modification intended to avoid complications when sampling from discrete populations.

What about the true significance levels for modified test procedures? As we mentioned earlier, when sampling from discrete populations (as evidenced by the presence of ties) the test is no longer distribution-free. However it turns out that, quite generally, tabulated significance levels furnish upper bounds for the true significance levels. The use of standard tables in connection with modified tests produces conservative results. The true descriptive significance level associated with the observed test result is at most as large as the tabulated value α'' for two-sided tests and α' for one-sided tests, or the corresponding probabilities obtained from the normal approximation. Actually for most practical purposes, the difference between the tabulated and true significance levels is negligible, unless a rather large proportion of all observations are tied at just a few values. In such a case a normal approximation incorporating a correction for ties may be helpful. Since this correction is rarely needed in practice, it is not given in this book.

90 When sampling from discrete populations there is not necessarily complete equivalence between confidence intervals and tests of hypotheses. Thus it may happen that either the lower or upper bound of a confidence interval for the population median η coincides with a hypothetical parameter value η_0. Acceptance or rejection of the hypothesis $\eta = \eta_0$ then depends on whether we use a closed or an open confidence interval. A closed confidence interval would accept the hypothesis $\eta = \eta_0$; an open interval would reject the hypothesis. The use of midranks in the computation of the Wilcoxon signed rank statistic T may be compared to taking a position somewhere between using an open confidence interval and using a closed one. It may happen, then, that use of a closed confidence interval, as we have recommended, suggests acceptance of the hypothesis $\eta = \eta_0$, while use of the Wilcoxon test based on midranks suggests rejection.

PROBLEMS

1 For each of the distributions in Figure 12.2, suggest a possible population.

2 Consider the following set of 8 observations: 5, 3, −1, 14, 7, −4, 13, 10.
   a. Find a confidence interval for the population median using two methods. Use a confidence coefficient close to .94.
   b. Test the hypothesis $\eta = 0$ against the alternative $\eta \neq 0$ using two different tests. What is the significance level of each of your tests? (If the two tests lead to different conclusions, how do you explain the difference?)
   c. Find a point estimate of the population median by two methods.

3 Ten students obtained the following grades on an examination: 72, 95, 79, 83, 93, 80, 91, 74, 70, 86. Test the hypothesis that the median score for this test is 75. Use two different tests.

4 Define some student population. Get the weights of 12 *randomly* selected students from this population. (For later use, also write down each student's height.) Find a confidence interval for the median weight of students in your population.

5 A testing company has found that 16 tires of a certain make provided the following number of miles of service:

27900	35100	29800	27700
26700	30700	26900	32400
24800	27400	24900	33300
31600	24300	28300	27600

Do these results support the claim that this kind of tire provides over 30000 miles of service on the average?

6 What method would you use to test each of the following hypotheses? In each case state the appropriate test statistic and critical region. What assumptions are necessary for the validity of your test?
   a. H_0: $\eta = 0$ against $\eta < 0$; $n = 30$, $\alpha = .10$.
   b. H_0: $\eta = 500$ against $\eta \neq 500$; $n = 20$, $\alpha = .01$.
   c. H_0: $\eta = 100$ against $\eta > 100$; $n = 10$, $\alpha = .05$.

7 The median score on a certain qualifying examination is known to be 75. It is claimed that "special" instruction increases a student's score on the examination. If the scores of 13 students who have had special instruction are available, explain how you would test the correctness of the claim. Be specific. State the hypothesis you would set up, the alternative, the appropriate test statistic and critical region, and any assumptions you would have to make.

8 In each of the following cases, assume that you are testing the hypothesis $\eta = \eta_0$ against the alternative $\eta \neq \eta_0$ and have computed the indicated value of the test statistic. What is the descriptive level of your result?
   a. $n = 40$, $S = 14$.
   b. $n = 16$, $T = 30$.
   c. $n = 16$, $T = 33$.
   d. $n = 100$, $S = 44$.
   e. $n = 50$, $T = 412$.

9 Solve Example 12.8 using the sign test (if appropriate).

10 Solve Example 12.6 using the Wilcoxon signed rank test (if appropriate).

*11 The confidence interval for the population median bounded by the second smallest and the second largest observations has confidence coefficient

$$\gamma = 1 - \frac{1 + n}{2^{n-1}}.$$

a. Check the appropriate entries in Table E.
b. Derive the formula for γ. (*Hint:* Generalize the derivation in Section 78.)

13

Comparative Experiments: Paired Observations

Comparative Experiments

91 In practice we often have to select one of several possible ways of performing a certain job. How can we find out which way is best? Or how can we determine whether there is so little difference between the various approaches that it is immaterial which one we choose? In this chapter and the next we discuss how to compare two competing methods. In Chapter 15 we take up the case of more than two methods. Here are a few examples.

Of two (or more) different ways of treating a certain disease, which one produces the best results? Is there any difference in the number of miles per gallon of gasoline that we can get from two (or more) competing makes of cars? Does high-test gasoline give better mileage than regular gasoline in cars that can use either? Is a new, supposedly superior, method suggested to replace an old established method for doing something really superior? For example, is the new mathematics we hear so much about these days really better than what has been taught under the name of mathematics until now? The reader should realize that problems like this last one are usually much more complex than may appear on the surface. We first must decide what we mean by new mathematics and old mathematics. Even if we agree on an answer to this question, what do we mean by saying that one method is better than the other? These are nonstatistical problems, but they have to be resolved before we can set up an experiment whose results are to be analyzed by statistical means. It is good to keep in mind that any problem must be properly formulated before statistical theory can successfully be applied to it.

Instead of worrying about new and old mathematics, let us consider something more agreeable, eight young ladies, for example.

92 *Lotion X versus Lotion Y*   Suppose we want to find out which of two suntan lotions, labeled *X* and *Y* for simplicity, provides better protection against sunburn. Let us also assume that 8 young ladies have volunteered to participate in an experiment in which they expose their backs to the sun for several hours, protected by suntan lotion, of course. This raises a problem of experimental design. How should the experiment be arranged so that it provides the most information about the question under investigation?

We could use lotion *X* on four of the young ladies and lotion *Y* on the other four. But a better procedure is to use both lotions on each of our young ladies, one lotion on the right side and the other lotion on the left side. By comparing only right and left sides of the *same* person, we eliminate variability among measurements caused by differences in skin sensitivity. However, one additional precaution is in order. For each person we have to determine, say, by means of the toss of a fair coin, which lotion goes on the left side and which lotion goes on the right side. Since in general the right and left sides of a person are not exposed to the sun in exactly the same way, we might introduce an unintentional bias by assigning, say, lotion *X* always to the right side. Randomization introduced by the toss of a coin avoids this and possibly other unsuspected biases. It also provides the basis for the statistical analysis that we are going to carry out.

After our young ladies have exposed their backs to the sun for the prescribed number of hours, we measure the degree of redness on each side. Here are the purely fictitious results:

X *Y*	*Y* *X*	*Y* *X*	*Y* *X*
51 46	45 48	53 52	48 62
+5	+3	−1	+14
X *Y*	*X* *Y*	*Y* *X*	*X* *Y*
64 57	51 55	42 55	60 50
+7	−4	+13	+10

For example, for our first young lady the toss of the coin determined that lotion *X* should go on the left side, leaving the right side for lotion *Y*. The accompanying numbers, 51 and 46, are measures of the degree of sunburn, the higher number indicating more severe sunburn. Finally we have the difference, +5, between the sunburn measurement for the side protected by lotion *X* and the side protected by lotion *Y*. For our remaining young ladies there is similar information. In all four, our coin assigned lotion *X* four times to the left side and four times to the right side. It just so happened that we obtained an even split. This is of course not always the case.

Since differences represent the most important information, we list them separately:

13.1   +5,   +3,   −1,   +14,   +7,   −4,   +13,   +10.

In each case the measurement for lotion Y has been subtracted from that for lotion X irrespective of whether lotion X was applied to the left side or the right side.

It is tempting to conclude merely on the basis of inspection of these differences that lotion Y provides better protection against sunburn than does lotion X. But we should remember our customary approach. Before we conclude that Y is really better, we want to be able to rule out the possibility that our result can be explained simply in terms of chance and that lotions X and Y are in fact equally efficient in preventing sunburn, or possibly equally inefficient. Thus we want to set up and test the null hypothesis that there is no difference in the protective ability of the two lotions. Only if our data indicate that this hypothesis should be rejected are we willing to conclude that lotion Y is superior to lotion X. First we must decide how we can translate the hypothesis to be tested into mathematical terms.

Let us have a closer look at our experiment. If there is no difference between the protective properties of the two suntan lotions, the labels X and Y are just that, labels and nothing else. Perhaps the two lotions smell different, but as far as sunburn protection is concerned, there is no difference between the two. Then differences between the two measurements of sunburn on the right and left are not caused by differences in protective ability of the two lotions, but are due to uncontrolled factors such as changing skin sensitivity and different exposure to the sun. In this case, a reversal of the result of the coin toss that determines which lotion goes on the right and which lotion goes on the left would not change the sunburn measurements one bit. But it would change our test results. A look at our first young lady shows how. We had the results

$$X \quad Y$$
$$51 \quad 46$$
$$+5$$

But if her coin had fallen the other way, we would have had the results

$$Y \quad X$$
$$51 \quad 46$$
$$-5$$

Thus if there is no difference in the protective ability of our two lotions, a reversal of the coin toss which assigns the two lotions changes a plus sign into a minus sign, and correspondingly, a minus sign into a plus sign. Since a fair coin falls heads or tails with probability $\frac{1}{2}$, it follows that plus and minus signs also have probability $\frac{1}{2}$. As a consequence, if there is no difference in the protective properties of the two lotions, a difference of -5 is just as likely to occur as a difference of $+5$. The same is of course true for any other number. In other words, when our hypothesis is correct, differences are symmetrically distributed about zero. In particular this means that the median difference is zero, $\eta = 0$.

93   In the preceding chapter we discussed two tests of the hypothesis $\eta = \eta_0$, the sign test and the Wilcoxon signed rank test. Because of the observed symmetry, both these tests are applicable in the present case with $\eta_0 = 0$. We shall illustrate both, though in practice just one would be chosen.

Since there are fewer negative than positive differences, the appropriate test statistic for the sign test is the number of negative differences, of which there are two. Checking Table E for the number of observed differences $n = 8$, we find that we cannot reject our hypothesis at any customary significance level.

In the present example it is appropriate to use a two-sided test, since at the start of the experiment we had no idea which lotion might be better. The situation would be different if it were known that lotion Y contains a "miracle" ingredient that is supposed to make it better than lotion X. In that case, we would test the hypothesis that lotions X and Y have equal protective ability against the alternative that Y provides better protection. We would then reject the null hypothesis only if the number of negative differences were sufficiently small.

We now turn to the Wilcoxon signed rank test. To compute the test statistic T, we need the ranks of the absolute differences:

difference	+5	+3	−1	+14	+7	−4	+13	+10
rank of absolute difference	4	2	1	8	5	3	7	6

We easily find $T = T_- = 1 + 3 = 4$. According to Table F we can reject the hypothesis $\eta = 0$ at significance level .055 since T is smaller than 5. The Wilcoxon signed rank test would seem to indicate that lotion Y does a better job than lotion X, provided we are willing to use a significance level as high as .055.

94   The reader is reminded of the remark made at the end of the preceding chapter. In practice, it is not appropriate to try one test procedure after another until one is found that rejects the hypothesis being tested. If an investigator wants to reject an hypothesis, it is always possible to find a test that will do so after looking at the data. But it would be much more honest to come right out and say so than to hide behind the screen of a statistical procedure.

It is instructive to investigate why, in the present example, the Wilcoxon signed rank test suggests rejection of the null hypothesis (at significance level .055) while the sign test does not. Looking at the data we observe that in the two cases where lotion X is associated with the less severe sunburn there is really very little difference between the measurements for lotions X and Y, just 1 and 4 points. On the other hand, in cases where lotion X is associated with more severe burns we notice such large differences as 10, 13, and 14. As far as the sign test is concerned, any negative difference carries as much weight as any positive difference. In particular the negative difference -4 counts as much as the positive difference $+14$. The same is not true of the Wilcoxon signed rank test. Associated with the difference $+14$ is the rank 8, while associated with the difference -4 is the much smaller rank 3. The sign test only observes that out of 8 observations 2 are negative and 6 are positive. The Wilcoxon signed rank test in addition observes that on the average the positive differences are considerably larger than the negative differences.

The reader should not interpret these remarks to imply that the sign test is not a useful test. They are only intended to bring out the differences between the two tests. In small samples the sign test conveys little information. But in large samples it often provides all the information that is needed to reach a sound conclusion. In addition, what the sign test lacks in sensitivity, it makes up in versatility. There are situations when only the sign test can be used. For example, suppose that in the suntan lotion problem we only had statements from each of our young ladies as to which side felt more sensitive from exposure to the sun, rather than actual measurements of degree of redness. As before, we can mark down a plus sign if the more sensitive side was protected by lotion X and a minus sign if it was protected by lotion Y, and then carry out the sign test. The sign test does not really require numerical information. Qualitative judgements are quite sufficient.

In many consumer preference studies, qualitative comparisons are the only ones obtainable. If you are asked to compare two different kinds of ice cream, you can presumably tell which kind you prefer, but you may not be able to give a meaningful numerical value for your degree of preference. In such cases the sign test offers the only possible method of analysis. In the suntan lotion example both tests are applicable. However, in view of the small number of observations available, most statisticians would presumably choose the Wilcoxon signed rank test.

The Analysis of Paired Observations

95 We are now ready to formulate our problem and its solution in general terms. But first we want to introduce some standard terminology. The problem of comparing two or more methods occurs so frequently that statisticians have developed a specialized terminology for it. Rather than speak of methods, they speak of *treatments*, even though in some cases no treatments in the everyday sense of the word may be involved. The hypothesis we are testing then states that the two or more treatments we are comparing have identical effects. Only if this hypothesis is rejected are we willing to concede that differences exist.

In this chapter we assume that we have two treatments and that we can obtain n pairs of observations, where one observation in each pair is from treatment 1 and the other from treatment 2. It is convenient to write these pairs of observations as $(x_1, y_1), \ldots, (x_n, y_n)$. Our analysis is then based on the n differences

$$d_1 = x_1 - y_1$$
$$\vdots$$
$$d_n = x_n - y_n.$$

Our discussion for the suntan lotion experiment shows that when the hypothesis of identical treatment effects is correct, the differences $d_1, \ldots, d_n$ have symmetric distributions with center of symmetry η at zero. On the other hand, when the two treatments have different effects, the differences $d_1, \ldots, d_n$ may be assumed to have distributions with median different from zero. We then have reduced our problem to one of testing the hypothesis $\eta = 0$ on the basis of n observed differences $d_1, \ldots, d_n$ against the two-sided alternative $\eta \neq 0$ or possibly the one-sided alternatives $\eta < 0$ or $\eta > 0$.

In Chapter 12 we saw that this hypothesis can be tested by either the sign test or the Wilcoxon signed rank test. Since the hypothetical value η_0 equals zero, the test statistics are particularly simple. Thus for the sign test, $S_- = \#(\text{negative differences})$ and $S_+ = \#(\text{positive differences})$. For the Wilcoxon signed rank test the absolute differences $|d_1|, \ldots, |d_n|$ are ranked from 1 to n, then T_- is set equal to the sum of ranks corresponding to negative d_i's, and T_+, the sum of ranks corresponding to positive d_i's. For both the sign and the Wilcoxon signed rank tests any pair of observations with $x_i = y_i$ is omitted from consideration (that is, differences of 0 are omitted) and the sample size is reduced correspondingly.

EXAMPLE 13.2

Suppose two methods of teaching reading are to be compared. It is claimed that method 2 is more effective than method 1, where the comparison is based on scores on some test measuring reading ability. The following experiment is conducted. Twenty first graders are divided into ten groups of two each in such a way that the two pupils in each group are as similar as possible with respect to previous reading experience, IQ, motivation, and other factors that may have a bearing on a student's ability to learn to read. Then in a random fashion one student in each group is assigned to reading method 1, the other to reading method 2. At the end of one year's instruction all 20 pupils take the same test. Suppose these are the results (the x-value in each pair is the score of the student taught by method 1):

TABLE
13.3

group	1	2	3	4	5	6	7	8	9	10
x	92	81	69	86	70	66	76	63	62	74
y	68	99	67	97	75	86	92	75	84	66
$d = x - y$	24	−18	2	−11	−5	−20	−16	−12	−22	8
rank	10	7	1	4	2	8	6	5	9	3

For illustrative purposes we again apply both the sign test and the Wilcoxon signed rank test. Since it has been claimed that method 2 is superior, a one-sided test against the alternative $\eta < 0$ is appropriate. The sign test statistic is $S_+ = 3$, which is not significant at levels given in Table E. We can find the appropriate descriptive level by using the normal approximation as illustrated in Chapter 12. Actually in the present case, Table A of the binomial distribution with $p = \frac{1}{2}$ provides the exact descriptive level: $P(S_+ \leq 3) = b(3) + b(2) + b(1) + b(0) = .117 + .044 + .010 + .001 = .172$.

For the Wilcoxon signed rank test we find $T_+ = 10 + 1 + 3 = 14$. According to Table F we again accept the null hypothesis. Our data do not justify the claim that method 2 is superior.

96   Confidence Intervals   In this chapter we have concentrated on the discussion of tests of hypotheses. It should be clear however that the two procedures for finding confidence intervals for a population median discussed in Chapter 12 can also be applied to the differences $d_i = x_i - y_i$ in this chapter to find a confidence interval for the median of all possible differences. Indeed in Problem 2a of Chapter 12 the reader was asked to find a confidence interval using as his data the 8 differences for the suntan lotion experiment. The resulting interval represents a confidence interval for the median difference in protection between lotions X and Y.

Example 13.2 was formulated in such a way that it required a yes or no answer to the question: Is method 2 superior to method 1 ? This formulation of the problem is appropriate if a decision has to be made whether method 2 should be used exclusively in the future. On the other hand when no such decision is contemplated, it may be much more informative to find a confidence interval for the median difference of scores to be achieved by the two methods. (See Problem 4.)

PROBLEMS   1 Consider the data in Table 15.5 comparing gasoline mileage for different makes of cars. Consider only makes C and F. Let η be the median difference in gasoline mileage between make C and make F.
   a. Test the hypothesis that $\eta = 0$.
   b. Find a confidence interval for η.
   c. Find a point estimate of η.

2 The pulse rates of 12 students before and after exercise were as follows:

before	62	65	71	58	75	68	60	72	75	91	59	68
after	80	79	96	87	87	91	90	92	96	89	81	84

Find confidence intervals (by two different methods) for
a. the median pulse rate before exercise,
b. the median increase in pulse rate due to exercise.

3 Fifty persons were asked which of two brands of soft drinks they preferred.
  a. If 18 stated that they preferred brand X, 26 stated that they preferred brand Y, and 6 stated that they did not like either, would it be fair to say that brand Y is preferred to brand X? Explain your reasoning.
  b. If the numbers are 19, 31, and 0 respectively, how would you answer the question?

4 Find an appropriate confidence interval for the data in Table 13.3. Does it contain zero?

14

Comparative Experiments: Two Independent Samples

97 In the preceding chapter we saw how to analyze comparative experiments in which measurements occur in pairs. For instance, each young lady in the suntan lotion experiment provided a measurement for lotion X *and* a measurement for lotion Y. Our analysis was based on the differences $x_i - y_i$. However, there are situations when it is impossible to use two different methods on the same test subject. If we want to compare the effectiveness of two different medical treatments, it generally is impossible to apply both to the same patient. In the problem of determining which of two methods of teaching first graders how to read is more successful, we cannot use both teaching methods on the same child. We got around the difficulty in this particular case by dividing the children participating in the experiment into groups of two in such a way that the two children forming a pair were as homogeneous as possible with respect to factors that were thought to have some bearing on a child's ability to learn how to read. By assigning one child in each group to method 1 and the other child to method 2 we obtained paired observations that could be analyzed according to the methods of Chapter 13. However, the success of this approach depends on how successful we are in forming homogeneous pairs. If no information is available or if no homogeneous pairs can be found, it is better not to attempt to use pairs as the basis of our experiment.

As an alternative, we simply divide the available test subjects into two groups. On the test subjects in the first group, we use method 1. On the test subjects in the second group, we use method 2. In this chapter we want to discuss methods for analyzing data resulting from such an experiment. But before we do so, some general remarks are in order.

It is usually best to keep the two experimental groups as nearly equal in number as possible. However, it is not always possible—or practical—to have exactly the same number of test subjects in each group. For this reason we want a method of analysis that is applicable whether the groups are of equal size or not.

In the suntan lotion experiment, we assigned lotions X and Y to the right or left sides of the backs of our young ladies at random, in order to avoid possible biases. For exactly the same reason, it is desirable to divide the available test subjects into two experimental groups in a random fashion. Many experiments have been spoiled from the very beginning by a nonrandom assignment of test subjects. If a subsequent analysis reveals group differences, there is always the possibility that the observed differences do not reflect differences between methods 1 and 2. They may simply be due to nonrandom assignments. In the reading experiment, suppose one school system uses method 1 and another school system uses method 2. At the close of first grade, the children in both school systems are given the same test and the children in group 2 get significantly higher scores than those in group 1. Does this mean that method 2 is superior to method 1? Not necessarily. The kindergarten preparation of the children in group 2 may have laid much more emphasis on preparation for reading than in group 1. There are other possible reasons. Only by assigning the available children to the two reading classes at random is it possible to avoid biases of this type.

Experimental conditions may not permit an effective randomization of observations corresponding to the two treatments to be compared. Thus consider the following example. Most Egyptian pyramids contain treasure chambers, practically all of which were discovered and plundered centuries ago. But no such treasure chamber has ever been found in the huge Chephren Pyramid. Recently scientists hypothesized that the possible existence of such a chamber in the Chephren Pyramid could be established by measuring the rate at which certain minute particles from outer space arrive in a vault under the pyramid. If there is an undiscovered chamber, the arrival rate of these particles would be higher than if the particles had to travel through solid rock. It is then necessary to compare measurements of arrival rates taken in a vault underneath the Chephren Pyramid with measurements of arrival rates of particles that are known to have traveled through solid rock. In an experiment like this it is very important for the experimenter to make certain that any potential experimental differences are due to treatment differences rather than extraneous factors.

Two Tests

98 We now turn to the problem of how to analyze two (independent) sets of observations from a comparative experiment. As before, we will present two different methods of analysis, but first let us look at a numerical example. Here are the scores that two different groups of pupils obtained on an examination (we shall find it convenient to refer to the first set as *x*-scores and to the second set as *y*-scores):

TABLE
14.1

group 1 (*x*-scores)	72	79	93	91	70	95	82	80	74	86
group 2 (*y*-scores)	78	66	65	84	69	73	71	75	68	90	76

We assume that pupils in groups 1 and 2 were taught by two different methods, and we want to find out whether or not the two methods of teaching are equally effective.

Even though there are only 21 observations in all, it is difficult to get a clear picture of what is going on by simply looking at these numbers. A graph is more enlightening. In Figure 14.2 the 21 scores are plotted along a horizontal axis, *x*-scores on top,

FIGURE
14.2

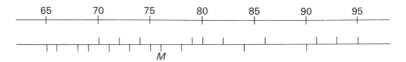

y-scores below. On the whole it would seem that pupils in group 1 have been doing better than pupils in group 2. But we should certainly know by now that such general impressions are insufficient to establish the superiority of teaching method 1 over teaching method 2. We need formal analysis to back up, or correct, our intuitive feeling. In particular, we want to be able to rule out the possibility that chance alone has brought about the configuration of results we have observed and that a repetition of the experiment could reverse the picture.

99   A formal comparison of *x*- and *y*-scores can take many different forms, resulting in different tests of the null hypothesis which states that there is no difference in the effectiveness of the two teaching methods. We look at two such tests.

The first test is extremely simple. We find the median M of all *x*- and *y*-scores. In our example, $M = 76$. We then divide the 21 scores into two groups. The first group consists of scores that are greater than the median (in symbols, $> M$). The second group consists of the remaining scores (in symbols, $\leq M$). Finally we find out how many *x*-scores and how many *y*-scores belong in each of these two groups. A two-way table like Table 14.3 is most convenient for tabulating the result.

TABLE
14.3

	$\leq M$	$> M$	totals
x-scores	3	7	10
y-scores	8	3	11
totals	11	10	21

When the null hypothesis is correct, the fact that a score is an x- or a y-score has no bearing on whether it falls above or below the median M of all scores. More precisely, if p_x is the probability that an x-score is greater than M and p_y, the probability that a y-score is greater than M, then the null hypothesis implies that $p_x = p_y$. But this hypothesis can be tested by means of the appropriate chi-square test of Chapter 11. We compute expected frequencies in the usual way. For example, the expected frequency in the upper left-hand corner is $E_{11} = \dfrac{11 \times 10}{21} = 5.24$. In this way we find that

$$\chi^2 = \frac{(3 - 5.24)^2}{5.24} + \frac{(7 - 4.76)^2}{4.76} + \frac{(8 - 5.76)^2}{5.76} + \frac{(3 - 5.24)^2}{5.24} = 3.84.$$

For one degree of freedom the descriptive level of significance associated with this value of χ^2 is .05. It would seem that too many x-scores are greater than M than can reasonably be expected under the null hypothesis. This test is known as the *median test*.

Like the sign test the median test simply counts how many observations are above or below the median. It does not pay attention to how much or how little a given observation differs from the median. The Wilcoxon test to be discussed now makes use of this latter kind of information. The chief recommendation of the median test is its simplicity both conceptually and computationally. Like the sign test it has its limitations, particularly when based on few observations. On the other hand it is quick and often does the job with a minimum of effort.

The median test is a test against the two-sided alternative that x-scores tend to be larger than y-scores or, vice versa, that y-scores tend to be larger than x-scores. It is possible to use the median test as a one-sided test. However the procedure is somewhat complicated and will not be discussed here. The Wilcoxon test to be discussed now can be used with equal ease against one- and two-sided alternatives.

100   The Wilcoxon test, or Wilcoxon-Mann-Whitney test as it is sometimes called, compares every x-score with every y-score counting the number U_x of times x-scores surpass y-scores, or alternatively, the number U_y of times y-scores surpass x-scores. From Figure 14.2 we see that the lowest x-score surpasses 4 y-scores, the next lowest x-score surpasses 5 y-scores, and so on. In this way we find that

$$U_x = 4 + 5 + 6 + 9 + 9 + 9 + 10 + 11 + 11 + 11 = 85.$$

The computation of U_y is even faster here,

$$U_y = 0 + 0 + 0 + 0 + 1 + 2 + 3 + 3 + 3 + 6 + 7 = 25.$$

(Note that $U_x + U_y = 110 =$ (number of x-scores) $\times$ (number of y-scores).)

It is rather obvious how U_x and U_y can be used to test the hypothesis that there is no difference between methods 1 and 2.

Under this hypothesis, x- and y-scores can be expected to be quite similar. But then U_x and U_y should be quite similar. On the other hand, when y-scores tend to be larger than x-scores, then U_x should be small; and if x-scores tend to be larger than y-scores, then U_y should be small. This suggests rejection of the null hypothesis when the smaller one of U_x and U_y is sufficiently small. As will be explained shortly, Table G indicates that our result suggests rejection at the .05-level.

The Wilcoxon Test

101 We shall now discuss the Wilcoxon test in general. We assume that we have two independent sets of observations, an x-set consisting of m observations

$$x_1, \ldots, x_m$$

and a y-set consisting of n observations

$$y_1, \ldots, y_n.$$

In the preceding example these are the two sets of test scores and, of course, $m = 10$ and $n = 11$. The hypothesis we want to test states that these two sets of observations have come from one and the same population. We are willing to believe that methods 1 and 2 produce different results only if our data indicate that the hypothesis that we have just stated should be rejected.

For the sake of our present discussion we assume that there are no ties among the observations, and, in particular, that no x-observation equals a y-observation. Required modifications when ties are present are discussed below.

As in the preceding example, we let

U_x = number of times x-observations are larger than y-observations,

U_y = number of times y-observations are larger than x-observations.

In order to compute U_x and U_y we have to compare every x-observation with every y-observation, mn comparisons in all. It then follows that $U_x + U_y = mn$, so that only one of the two quantities has to be computed from the given data. The other can always be obtained by subtraction. In the earlier example, inspection shows that the computation of U_y requires much less time than the computation of U_x. However, one advantage of computing both quantities from the data is that we can check the correctness of our computations by checking that $U_x + U_y = mn$.

Let U equal the smaller one of U_x and U_y. The Wilcoxon test rejects the hypothesis that the x- and y-samples have come from the same population if U is *smaller* than d, where d is listed in Table G. This test has significance level $\alpha'' = 1 - \gamma$ and is a test against the two-sided alternative which states that, on the average, x-observations are greater than y-observations or, vice versa, that y-observations are greater than x-observations. For a one-sided test with significance level $\alpha' = (1 - \gamma)/2$, we use the test statistic U_x if, under the alternative hypothesis, x-observations are supposed to be smaller than y-observations and the test statistic U_y if, under the alternative hypothesis, y-observations are supposed to be smaller than x-observations. (See Problem 2.)

Table G is arranged according to sample sizes m and n. In order to save space it is assumed that if $m \neq n$, m is the smaller of the two sample sizes. Since x and y are simply symbols for the two sets of observations, it is no restriction to use $x_1, \ldots, x_m$ for the set with fewer observations. Table G provides information for $3 \leq m \leq n \leq 12$.

EXAMPLE 14.4a   If $m = 10$ and $n = 11$, for a two-sided test we reject the null hypothesis at significance level .051 when $U < d = 28$. For the data in Table 14.1, $U = 25$. Thus rejection of the null hypothesis at the .05 level is clearly indicated.

102   ***Distribution of U***   In general, we do not derive the sampling distributions that underlie given test and estimation procedures. However, one advantage of the nonparametric approach is the ease with which the sampling distribution of a test statistic like U can be found, at least for small samples. We shall illustrate this point for the case $m = 2, n = 3$.

Since the values of the statistics U_x and U_y depend only on the position of x-observations relative to y-observations, we need only enumerate all possible arrangements of two x's and three y's and compute the appropriate value of U for each. Thus we may write

$$xxyyy$$

to indicate that both x-observations are smaller than all three y-observations. For this arrangement, clearly $U_x = 0$, $U_y = 6$ and therefore $U = 0$. In all, there are 10 different arrangements of two x's and three y's. Table 14.5 lists all 10 arrangements together with the corresponding values of U_x, U_y, and U. Since the null

TABLE 14.5

Values of U; $m = 2, n = 3$

arrangement	U_x	U_y	U
xxyyy	0	6	0
xyxyy	1	5	1
xyyxy	2	4	2
xyyyx	3	3	3
yxxyy	2	4	2
yxyxy	3	3	3
yxyyx	4	2	2
yyxxy	4	2	2
yyxyx	5	1	1
yyyxx	6	0	0

hypothesis states that the x- and y-observations come from one and the same population so that x's and y's are meaningless labels, all 10 arrangements have the same probability, namely $\frac{1}{10}$. We then easily find the distribution for U given in Table 14.6. In particular, the test that rejects the null hypothesis when $U = 0$

TABLE 14.6

Distribution of U; $m = 2$, $n = 3$

$$P(U = 0) = \tfrac{2}{10}$$
$$P(U = 1) = \tfrac{2}{10}$$
$$P(U = 2) = \tfrac{4}{10}$$
$$P(U = 3) = \tfrac{2}{10}$$

has significance level .20. The corresponding entry in Table G would read:

$$d = 1 \qquad \gamma = .80 \qquad \alpha'' = .20 \qquad \alpha' = .10.$$

103 ***Normal Approximation*** As on previous occasions, unless m or n is quite small, say less than 8, approximate probabilities relating to the U-test can be obtained with the help of the normal distribution. The appropriate values of μ and σ are $mn/2$ and $\sqrt{mn(m + n + 1)/12}$.

EXAMPLE 14.4b

What is the descriptive level of significance associated with the computed value $U = 25$ for our numerical example? We need

$$P(U \leq 25) = \phi\left(\frac{25 + \tfrac{1}{2} - 55}{\sqrt{10 \times 11 \times 22/12}}\right) = \phi(-2.08) = .019.$$

It follows that the descriptive significance level is approximately $2 \times .019 = .038$.

104 ***Tied Observations*** Our preceding discussion of the U-test is based on the assumption that there are no ties among the observations. Since ties do occur in practice, we shall now indicate how to deal with them. First of all, ties among x- and y-observations require a redefinition of U_x and U_y. According to our earlier instructions, to find the value of U_x we compare every x-observation with every y-observation and count 1 whenever the x-observation is greater than the y-observation, and 0 when it is less. We now add the following rule. Whenever an x-observation equals a y-observation, we count $\frac{1}{2}$. A corresponding change applies to the computation of U_y. This additional rule does not change the relationship $U_x + U_y = mn$, since a tie between an x- and a y-observation adds $\frac{1}{2}$ to both U_x and U_y. Even in the presence of ties it is sufficient to compute one of the two statistics and find the other by subtraction. The remarks about the true significance level of the Wilcoxon signed rank test in the presence of tied observations (Section 89) apply also in this case.

105 *Alternative Form of Wilcoxon Test* Wilcoxon actually proposed a statistic other than U for carrying out the test that bears his name. To find this other statistic we arrange the $m + n$ x- and y-observations according to size and then replace the observations by their ranks. Let R_x be the sum of the ranks that correspond to x-observations, and R_y, the sum of the ranks corresponding to y-observations. Wilcoxon suggested using R_y as the test statistic and rejecting the null hypothesis whenever R_y is sufficiently large or sufficiently small. Of course, this test requires its own appropriate tables.

We want to show now that the four quantities $U_x, U_y, R_x,$ and R_y are very closely related. If we know any one of the four, we can easily compute the other three. We know already that $U_x + U_y = mn$, allowing us to compute one from the other. It follows from the above definitions that

$$R_x + R_y = 1 + \cdots + (m + n) = (m + n)(m + n + 1)/2,$$

allowing us to compute one rank sum from the other. What we still need is a relationship between U_x and R_x or U_y and R_y. An example will show the nature of such a relationship.

We consider the following two samples:

$$\begin{array}{lllll} \text{x-sample:} & 69 & 61 & 76 & (m = 3) \\ \text{y-sample:} & 73 & 85 & 79 & 72 \quad (n = 4) \end{array}$$

Let us start by ordering all 7 observations according to size and computing U_x and U_y in the usual way:

observation	61	69	72	73	76	79	85	
x or y	x	x	y	y	x	y	y	
number of times x surpasses y	0	0	–	–	2	–	–	$U_x = 2$
number of times y surpasses x	–	–	2	2	–	3	3	$U_y = 10$

Next we add 1 for the first x-observation, 2 for the second x-observation, and so on; and then do the same for the y-observations:

observation	61	69	72	73	76	79	85
x or y	x	x	y	y	x	y	y
scores for computing U_x or U_y	0	0	2	2	2	3	3
add to x-score	1	2	–	–	3	–	–
add to y-score	–	–	1	2	–	3	4
result of addition	1	2	3	4	5	6	7

The resulting numbers are seen to be the ranks of the x- and y-observations when we rank all 7 observations from the smallest to the largest. In particular, $R_x = 1 + 2 + 5 = 8 = U_x + 1 + 2 + 3$ and $R_y = 3 + 4 + 6 + 7 = 20 = U_y + 1 + 2 + 3 + 4$. For any m and n we have the following formulas (see Problem 10):

$$R_x = U_x + m(m + 1)/2,$$
$$R_y = U_y + n(n + 1)/2.$$

There are two reasons for mentioning this second form of the Wilcoxon test. Our formulas allow us to compute U_x (or U_y) from R_x (or R_y). More importantly however, in Chapter 15 we discuss how to compare more than two treatments. It is simpler to generalize from the Wilcoxon test using rank sums R_x and R_y than from the test based on statistics U_x and U_y.

EXAMPLE 14.7

A car owner wants to know whether he should believe the advertising claim that high-test gasoline improves gas mileage, so he conducts the following experiment. He repeatedly observes the number of miles he can drive with 10 gallons of gasoline. Here are his results:

high-test gasoline (x-scores)	175	139	186	228	211	182	185
regular gasoline (y-scores)	186	206	149	213	156	181	180

In order to get an answer to his question, he tests the hypothesis that all 14 observations have come from the same population against the alternative that (in his car) high-test gasoline produces better gas mileage than regular gasoline. The appropriate test statistic is U_y. We shall compute U_y according to the formula $U_y = R_y - n(n + 1)/2$. The following arrangement of the observations gives us the value of R_y:

observation	139	149	156	175	180	181	182	185	186	186	206	211	213	228
x or y	x	y	y	x	y	y	x	x	x	y	y	x	y	x
rank	1	2	3	4	5	6	7	8	9.5	9.5	11	12	13	14

(Since an x- and a y-observation are tied for 9th and 10th place, both receive the midrank 9.5.) We then find $R_y = 2 + 3 + 5 + 6 + 9.5 + 11 + 13 = 49.5$ and $U_y = 49.5 - 28 = 21.5$. According to Table G, this result is highly insignificant. There is no evidence whatsoever that high-test gasoline provides better gas mileage than regular gasoline.

Confidence Intervals for a Shift Parameter

106 In this chapter as in Chapter 13, we have primarily concerned ourselves with tests of hypotheses. However, experience shows that observations from two comparable, though different, treatments often follow distributions that are similar in shape but shifted relative to one another as in Figure 14.8. If this seems a reasonable model, we may then try to find a confidence interval for the amount of shift Δ of the x-population relative to the y-population.

FIGURE
14.8

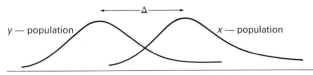

Such a confidence interval is easily found. Let $v_{ij} = x_i - y_j$, the difference between the ith x-observation and jth y-observation. Each of the mn differences corresponding to subscripts $i = 1, \ldots, m; j = 1, \ldots, n$ can be considered a point estimate of Δ. A confidence interval for Δ is bounded by the dth smallest and dth largest of all these "estimates", where d is read from Table G.

107 There is a close relationship between this confidence interval and the Wilcoxon test discussed earlier in this chapter. If the shift parameter Δ in Figure 14.8 is zero, the x- and y-observations come from the same population, so the null hypothesis tested by the Wilcoxon test is satisfied. Our confidence interval contains the parameter value $\Delta = 0$ if and only if the Wilcoxon test accepts the hypothesis that the x- and y-observations have come from the same population. Thus as on earlier occasions, the confidence interval can be used in place of a test of a hypothesis.

As an illustration, let us find a confidence interval for the shift parameter Δ using the data in Table 14.1. For $m = 10, n = 11$ we find from Table G that the value $d = 20$ corresponds to a confidence interval with confidence coefficient .99. For such a confidence interval we need the 20th smallest and the 20th largest of the 110 possible differences $v_{ij} = x_i - y_j$. We may proceed as follows. We arrange the x-observations from the smallest to the largest and the y-observations from the largest to the smallest:

x: 70 72 74 79 80 82 86 91 93 95
y: 90 84 78 76 75 73 71 69 68 66 65

We obtain small differences v_{ij} by subtracting large y's from small x's: $70 - 90 = -20, 72 - 90 = -18$, etc. until we find the 20th difference -3. (Note that equal differences like $72 - 76 = 74 - 78 = 80 - 84 = 86 - 90$ are counted as often as they occur.) Similarly, large differences v_{ij} are obtained by subtracting small y's from large x's: $95 - 65 = 30, 95 - 66 = 29$, etc., until we find the 20th difference from the top, $+20$. Thus we have the confidence interval

$$-3 \leq \Delta \leq 20.$$

The median score of students taught by method 1 may be as much as 20 points higher than that of students taught by method 2. But it may also be as much as 3 points lower.

We note that the confidence interval does contain the value $\Delta = 0$. This is no contradiction to our earlier test result. The confidence interval on page 139 provides a test with significance level $\alpha'' = 1 - \gamma = 1 - .99 = .01$. The earlier test indicated that the null hypothesis could be rejected at significance level .05, but not at the level .01. Indeed the normal approximation indicated a descriptive level of about .04.

108 *Graphical Determination* A systematic ordering of the v_{ij} according to size can be achieved graphically as follows. On a sheet of graph paper we mark $x_1, \ldots, x_m$ along the horizontal axis, and $y_1, \ldots, y_n$ along the vertical axis. We then plot the mn points with coordinates (x_i, y_j). These points represent the differences v_{ij}. They are ordered according to increasing (decreasing) values $v_{ij} = x_i - y_j$ by sliding a 45°-line from the upper left (lower right) and counting points.

The necessary steps for the graphic solution of the earlier example are indicated in Figure 14.9. Sliding a 45°-line from the upper left we find that the 20th point has coordinates (70, 73) or (72, 75), both of which give $v_{ij} = -3$. There are four choices for the 20th point when sliding a 45°-line from the lower right. We can choose any one of these four points, for example, the point (95, 75), giving the value $v_{ij} = 95 - 75 = 20$.

FIGURE 14.9

Graphical Determination of Shift Parameter Estimates

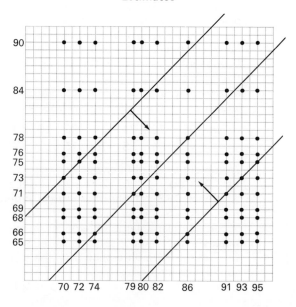

109   *Point Estimates of* Δ We have observed that every difference $x_i - y_j$ furnishes a point estimate of Δ. The most appropriate of all mn estimates is the one in the middle when we arrange all mn estimates from the smallest to the largest, that is, the median of the v_{ij}'s.

For our example there are 110 v_{ij}'s. Therefore the median is halfway between the 55th and 56th v_{ij}. According to Figure 14.9 both the 55th and 56th points fall on the same line. One of the points has coordinates (79, 71), so that our point estimate of Δ is $79 - 71 = 8$.

PROBLEMS

1 For the Chephren Pyramid example,
   a. formulate the null hypothesis to be tested;
   b. specify the alternative hypothesis.

2 Justify the choice of test statistic specified on page 135 for one-sided alternatives.

3 For the data of Example 14.7 compute U_y directly without using R_y. Which method of computation seems more convenient?

4 Consider the data in Table 15.1a giving the grades of three groups of pupils. Using the data for groups 2 and 3 only,
   a. test the hypothesis that the two methods of instruction do not produce different results;
   b. find a confidence interval for a possible difference in test results.

5 A random sample of 9 families in community A reported the following yearly incomes: 9100; 10300; 6800; 13500; 5500; 7900; 11000; 14000; 7200. A random sample of 8 families in community B reported the following yearly incomes: 11500; 8300; 6100; 19000; 14500; 7200; 10500; 9400.
   a. Would you say that the two communities differ with respect to family income?
   b. Find a confidence interval for the difference in family income for the two communities.
   c. In what sense do the answers under (a) and (b) agree?

6 It is claimed that the nicotine content of cigarettes of brand Y is lower than that of cigarettes of brand X. A laboratory makes the following determination of nicotine content (in milligrams):

   Brand X: 20, 23, 25, 21, 26
   Brand Y: 18, 22, 24, 19, 20

   Do you agree with the claim? Why?

7 Find R_x and R_y for the data in Table 14.1.

8 On 8 occasions of cloud seeding the following amounts of rainfall were observed: .74, .54, 1.25, .27, .76, 1.01, .49, .70. On 6 control occasions (when no cloud seeding took place), the following amounts of rainfall were measured: .25, .36, .42, .16, .59, .66. Would you feel justified in claiming that cloud seeding increases amount of rainfall?

*9 Prove that $R_x + R_y = (m + n)(m + n + 1)/2$ even when ties are present.

*10 Prove that $R_x = U_x + m(m + 1)/2$ and $R_y = U_y + n(n + 1)/2$.

*11 Find the distribution of U for $m = n = 3$ and check the entries in Table G.

15

Comparative Experiments: *k* Samples

110   In preceding chapters we considered the problem of how to compare two different methods or treatments and how to decide which, if either, is better. In practice, more than two methods are often available. We now want to generalize our test procedures to cover the comparison of any number of methods. For example, in evaluating the effectiveness of programmed instruction in some such course like high school algebra, we may give the same examination to three different groups of students:

> Students who have had programmed instruction but no supplementary discussions.
>
> Students who have had programmed instruction and in addition have had the opportunity to discuss the material with a qualified instructor.
>
> Students who have had regular classroom instruction but have not used programmed materials.

By considering, in addition, instruction by TV we could easily enlarge the scope of the experiment, ending up with considerably more than 3 groups of students.

The Kruskal-Wallis Test

111   Let us look at a set of examination scores involving 3 groups of students. In order to keep the arithmetic simple we shall assume that there are only 7 pupils in each group. As we shall see later, our method of analysis can be used for any number of groups, where the number of observations is not necessarily the same for each group. Thus in an evaluation of the effectiveness of various methods of teaching high school algebra, we may start with an

143

equal number of pupils in each group, but in the course of the school year some pupils may drop out for one reason or another. It would be a great waste of effort, time, and money if the results for the remaining pupils could not be used to compare the various teaching methods.

Our basic data are in Table 15.1a. We are fortunate not to have to worry about drop-outs. As in the alternate approach to the

TABLE 15.1a

Grades for Three Groups of Students

Group 1	Group 2	Group 3
66	77	79
68	68	83
80	67	78
72	85	92
74	89	98
57	74	69
75	61	91

Wilcoxon test we replace observations by ranks. In other words, we replace the smallest among the 21 observations by 1, the next smallest by 2, and so on, using midranks where ties occur. The results of the rankings, together with the sum of ranks for each group, are given in Table 15.1b. The hypothesis that the teaching methods do not differ in their effectiveness (as measured by the results of the examination) is rejected if the sums of the ranks for the three groups differ sufficiently from each other.

TABLE 15.1b

Group 1	Group 2	Group 3
3	12	14
5.5	5.5	16
15	4	13
8	17	20
9.5	18	21
1	9.5	7
11	2	19
53	68	110

For the Wilcoxon test it is possible to look in an appropriate table and decide whether the observed rank sums suggest rejection or acceptance of the hypothesis being tested. For more than two rank sums the corresponding tables are very awkward, so we rely on something along the lines of a chi-square test. In our example, the ranks range from 1 to 21, so that the average rank is 11. Since each rank sum contains 7 terms, if there is no difference among teaching methods each rank sum has expectation 77. A suitable quantity for measuring deviation from expectation is given by the sum of squares

$$(53 - 77)^2 + (68 - 77)^2 + (110 - 77)^2 = 1746.$$

While theoretically it would be possible to have a table from which we could find out whether to accept or reject our hypothesis, we can make use of the standard chi-square distribution with two degrees of freedom simply by multiplying the sum of squares by a suitable constant, as we shall see presently. We shall see that the hypothesis that the three teaching methods produce identical results can be rejected at the .05 level. It would appear that method 3 produces higher test scores than either method 1 or 2.

112 We now formulate the problem and its solution in general terms. First we define our notation. The number of treatments to be compared is k, where k is a number greater than 2. There are n_i observations for the ith method, $i = 1, \ldots, k$. The total number of observations is $N = n_1 + \cdots + n_k$. The sum of the ranks for the ith method is denoted by R_i. The test of the null hypothesis that there exist no differences among the k methods is based on the statistic

$$H = \frac{12}{N(N+1)} \left[\frac{1}{n_1} \left(R_1 - n_1 \frac{N+1}{2} \right)^2 + \cdots + \frac{1}{n_k} \left(R_k - n_k \frac{N+1}{2} \right)^2 \right]$$

which is more easily computed in the form

$$H = \frac{12}{N(N+1)} \left[\frac{R_1^2}{n_1} + \cdots + \frac{R_k^2}{n_k} \right] - 3(N+1).$$

The hypothesis being tested is rejected if the value of H is sufficiently large according to the chi-square distribution with $k - 1$ degrees of freedom.

The formula for H is particularly simple when $n_1 = \cdots = n_k = n$. Then

$$H = \frac{12}{nN(N+1)} [R_1^2 + \cdots + R_k^2] - 3(N+1).$$

This test is known as the Kruskal-Wallis test.

EXAMPLE 15.2a For the data in Table 15.1a we find (using the rank sums in Table 15.1b),

$$H = \frac{12}{7 \times 21 \times 22} (53^2 + 68^2 + 110^2) - 3 \times 22 = 6.5.$$

For 2 degrees of freedom and significance level .05 Table D gives the critical value 5.99. Since the observed value of H surpasses this critical value, the null hypothesis that the three teaching methods are equally effective should be rejected, as we indicated earlier.

113 Let us look at one further application of the Kruskal-Wallis test. A testing company compared five brands of automobile tires with respect to the distance required to bring a car traveling at a given speed on wet pavement to a full stop after locking the brakes. Table 15.3a gives the observed stopping distances in feet.

TABLE 15.3a

Braking Distances for Five Brands of Tires

brand of tire

A	B	C	D	E
151	157	135	147	146
143	158	146	174	171
159	150	142	179	167
152	142	129	163	145
156	140	139	148	147
			165	166

In order to test the hypothesis that no differences exist, we convert the $N = 27$ observations to ranks and sum the ranks for each brand. This is done in Table 15.3b. The computation of the test

TABLE 15.3b

brand of tire

A	B	C	D	E
15	18	2	11.5	9.5
7	19	9.5	26	25
20	14	5.5	27	24
16	5.5	1	21	8
17	4	3	13	11.5
			22	23
75	60.5	21	120.5	101

statistic H is now straight-forward:

$$H = \frac{12}{27 \times 28}\left(\frac{75^2}{5} + \frac{60.5^2}{5} + \frac{21^2}{5} + \frac{120.5^2}{6} + \frac{101^2}{6}\right) - 3 \times 28 = 12.3.$$

This value of H surpasses the .025 critical value of chi-square with 4 degrees of freedom. Thus we may reject the null hypothesis. The braking distance is not the same for the different brands of tires.

In practice, we hardly would want to stop our analysis at this point. Rather, we would want to go on and find out which brands differ appreciably from the others in braking ability. The original data indicate that any differences will not be large. But when it comes to braking ability, even a few feet may spell the difference between an accident and a safe ride. There is not much doubt that brand C is better than brand D. But what about brands B and C or A and C? Suppose brand A is considerably cheaper than brand C. Should a car owner in need of new tires buy brand A or brand C tires (assuming that braking ability is the deciding factor)? The *method of multiple comparisons* provides answers to this and similar questions.

114   ***Multiple Comparisons***   When comparing several treatments the method of multiple comparisons allows us to decide which treatments differ from which (if any). For $i = 1, \ldots, k$, we denote the sum of the ranks corresponding to the ith treatment by R_i and let $r_i = R_i/n_i$ denote the average rank. Let $j = 1, \ldots, k$ and $i \neq j$. Intuitively, we would say that treatments i and j do not differ if $r_i - r_j$ is sufficiently close to zero. On the other hand, if $r_i - r_j$ takes a sufficiently large positive value, we would say that treatment i produces larger measurements than treatment j, with a corresponding statement for large negative differences.

To make these statements more precise, we choose a significance level α such that we are willing to tolerate probability at most α of declaring that two or more of the k treatments differ, when in fact all k treatments are identical. When we use the Kruskal-Wallis test, we carry out just one overall comparison of all k treatments. It is then usually appropriate to use a rather small significance level like .05 or even .01. However, in the present approach we are interested in a large number of pairwise comparisons. Indeed, if we compare every treatment with every other treatment, the total number of possible comparisons is $k(k - 1)/2$, a number tabulated in Table 15.4 for k from 2 to 10. It is then reasonable to tolerate a considerably larger probability of making at least one false decision that two given treatments differ when in fact they do not. Thus we may very well choose α as large as .20 or even .25.

TABLE 15.4

Number of Pairwise Comparisons

k	2	3	4	5	6	7	8	9	10
$k(k - 1)/2$	1	3	6	10	15	21	28	36	45

For a given value α we can achieve our aim of having probability at most α of declaring that two or more treatments differ when in fact all k treatments are identical in the following way. In Table C (or by interpolation in Table B) we find the value z that corresponds to the upper tail probability $\alpha' = \alpha/k(k - 1)$. For every pair i, j, with $i \neq j$, we compute

$$z_{ij} = \frac{r_i - r_j}{\sigma_{ij}}$$

where

$$\sigma_{ij} = \sqrt{\frac{N(N + 1)}{12} \left(\frac{1}{n_i} + \frac{1}{n_j} \right)}$$

$$= \sqrt{\frac{k(N + 1)}{6}} \text{ , if } n_1 = \cdots = n_k.$$

Finally, we declare that on the average, treatment i produces smaller measurements than treatment j if $z_{ij} < -z$; larger measurements if $z_{ij} > z$; and similar measurements if $-z \leq z_{ij} \leq z$.

Let us return now to the tire problem where we wanted to compare five brands of tires pair by pair with respect to braking ability. Suppose that we are willing to tolerate an α-probability of

.25. Then $\alpha/k(k-1) = \frac{.25}{20} = .0125$ and $z = 2.241$. We first compute the rank averages $r_i = R_i/n_i$, where the rank sums R_i are taken from Table 15.3b:

brand	A	B	C	D	E
rank sum	75	60.5	21	120.5	101
sample size	5	5	5	6	6
rank average	15.0	12.1	4.2	20.1	16.8

Our earlier instructions were to compare ratios $(r_i - r_j)/\sigma_{ij}$ with the critical z-value, 2.241. An equivalent, and in the present case more convenient procedure is to compare differences $r_i - r_j$ with the product $z\sigma_{ij}$. There are three different σ_{ij}'s depending on the sample sizes involved. Thus we find:

sample sizes	σ_{ij}	$z\sigma_{ij}$
5 and 5	5.02	11.25
5 and 6	4.81	10.78
6 and 6	4.58	10.26

We can now say that two brands differ in their performance, if the difference in rank averages surpasses the appropriate value for $z\sigma_{ij}$. It follows that brand E can be expected to require larger stopping distances than brand C, since the rank difference $16.8 - 4.2 = 12.6$ surpasses the appropriate factor 10.78. Similarly, brand D can be expected to require a longer stopping distance than brand C. However these are the only two pairs that differ significantly in their performance, since every other difference $r_i - r_j$ is numerically smaller than the appropriate factor $z\sigma_{ij}$.

The Friedman Test

115 In actual experiments we often obtain data that look as if they could be analyzed by means of the Kruskal-Wallis test. Here we have the average number of miles per gallon of gasoline that cars built by three different manufacturers achieved in a well-publicized economy run:

TABLE
15.5

Gasoline Mileage for Various Cars

model	manufacturer		
	G	F	C
compacts	20.3	25.6	24.0
intermediate 6's	21.2	24.7	23.1
intermediate 8's	18.2	19.3	20.6
full size 8's	18.6	19.3	19.8
sports cars	18.5	20.7	21.4

We should like to have an answer to the following question: Do the three manufacturers achieve any consistent differences in gasoline mileage irrespective of car model? It may appear as though we have the same kind of data as in the example involving the comparison of the three methods of teaching algebra. There are again three treatments—the three manufacturers of cars. But outward appearance is deceptive. It would be a serious mistake to analyze the above data by the H-test. Let us see why.

In the present case our observations appear in distinct groups of three, one group for each model class. This is quite different from the setup of Example 15.1a, where pupils were not matched in groups. What we are considering are generalizations of the two different two-sample setups discussed in Chapters 13 and 14. The Kruskal-Wallis test generalizes the Wilcoxon test of Chapter 14 for independent samples. What we need now is a test procedure that is appropriate when observations are collected in groups of k, each group containing exactly one observation for each of the k treatments under investigation. This is a generalization of the setup of Chapter 13, when observations occurred in pairs. Since observations belonging to the same group are generally related, the assumption of independent samples underlying the Kruskal-Wallis test is likely to be violated.

For the Kruskal-Wallis test we assign ranks within the set of all observations, in order to compare every observation with every other observation. Under the present setup it is appropriate to assign ranks separately within each group. (It certainly makes little sense to compare gasoline mileage of compacts with that of full size V8's.) We then find the following rankings:

	G	F	C
compacts	1	3	2
intermediate 6's	1	3	2
intermediate 8's	1	2	3
full size 8's	1	2	3
sports cars	1	2	3
	$R_1 = 5$	$R_2 = 12$	$R_3 = 13$

As in the case of the Kruskal-Wallis test we compute the sum of ranks for each treatment. However since we have used a different ranking procedure, these rank sums have to be analyzed differently. for k treatments and n groups the appropriate test statistic is

$$Q = \frac{12}{nk(k+1)} \{ [R_1 - \tfrac{1}{2}n(k+1)]^2 + \cdots + [R_k - \tfrac{1}{2}n(k+1)]^2 \}$$

$$= \frac{12}{nk(k+1)} [R_1^2 + \cdots + R_k^2] - 3n(k+1).$$

The hypothesis that there are no differences among the k treatments is rejected if the value of Q surpasses the tabulated value of chi-square with $k - 1$ degrees of freedom at the chosen significance level.

For the gasoline mileage data, $k = 3$ and $n = 5$, so that

$$Q = \frac{12}{5 \times 3 \times 4} (5^2 + 12^2 + 13^2) - 3 \times 5 \times 4 = 7.6,$$

a value that indicates rejection of the null hypothesis at the .02 level. It would seem that the cars produced by manufacturer G deliver lower gasoline mileage than the cars produced by manufacturers F and C.

116 *Multiple Comparisons*   Our intuitive judgment is borne out by the method of multiple comparisons. For the setup of the Friedman test we compare rank sum differences $R_i - R_j$ with $\pm z\sigma$, where $\sigma = \sqrt{nk(k + 1)/6}$.

Suppose we choose $\alpha = .10$ for our gasoline mileage data. Then $\alpha/k(k - 1) = \frac{.10}{6} = .0167$ and therefore $z = 2.128$. Further, $\sigma = \sqrt{5 \times 3 \times 4/6} = 3.162$, so $z\sigma = 6.73$. Since both $R_2 - R_1$ and $R_3 - R_1$ are greater than 6.73, we conclude that the gasoline mileage of cars manufactured by G is lower than that of cars manufactured by either F or C.

Completely Randomized and Randomized Block Designs

117 We have discussed two different tests involving the comparison of k treatments. The two tests are not interchangeable. Which of the two tests is applicable in a given situation, if either, depends on the way the experiment that produces the basic data is designed. Suppose we have N experimental units available for our experiment. In the teaching experiment the 21 pupils are the experimental units. In the gasoline mileage experiment the 15 cars are the experimental units. The Kruskal-Wallis test is appropriate whenever the N available experimental units are assigned completely randomly to the k treatments, n_1 to treatment 1, n_2 to treatment 2, . . . , n_k to treatment k. We then have a *completely randomized design*.

On the other hand, it often happens in practice that the available $N = kn$ experimental units are, or can be, divided into n homogeneous groups, or *blocks* as we shall say from now on, each block consisting of k units. In such a case, all k treatments should be used exactly once (in random order) in each of the n blocks, resulting in a so-called *randomized block design*. Such data are analyzed by the Friedman procedure.

The second experimental setup calls for some further comments. In many experiments blocks arise in a natural way. Thus in the gasoline mileage example the division of experimental units into blocks is imposed by the various model classes. But in other experiments the experimenter may have to make a conscious effort to form appropriate blocks (see Problem 6). The problem of whether to form blocks or not is discussed in detail in courses on the design of experiments. Here we limit ourselves to one or two simple comments. Clearly a block design of the type we have discussed requires that the number of experimental units be a multiple of the number of treatments. However, this does not mean that whenever this condition is satisfied, we should try to arrange for blocks. It is advisable to use a block design only if, within blocks, greater homogeneity of the factors having a bearing on the phenomenon being investigated can be achieved.

118 *The Problem of n Rankings* The Friedman test is often used to solve a problem known as the *problem of n rankings*.* In this problem, n persons, often referred to as *judges*, are asked to rank k objects (like contestants for a prize, brands of consumer goods, and so on) in order of preference. Here, the judges correspond to blocks; the objects, to treatments. The purpose of such an experiment is to find out whether there is some degree of agreement among the n judges with respect to their order of preference. (In this sense the problem of n rankings has some similarity with the topic to be discussed in Chapter 16.) Agreement among the judges is indicated by a high value of Q. On the other hand, a low value of Q may mean one of two things: pronounced disagreement among the judges, or random arrangement of the objects due to a lack of preference. This second possibility corresponds to the earlier null hypothesis that k treatments do not produce different effects. If the objects to be ranked by the judges differ little, the judges are likely to arrange them in random order.

Paired Comparisons

119 When the number k of treatments that we want to compare is large, it may happen that the only kind of homogeneous blocks that can be found contain fewer than k units. This has lead to the development of so-called *incomplete block designs*. We consider only one very simple such design and its analysis.

We have mentioned consumer preference ratings as an application of the problem of n rankings. Experience shows that consumers provide the most reliable information if each consumer is asked to compare only two brands. A consumer participating in such an experiment represents one block with two experimental units. The entries in Table 15.4 can then also be interpreted as the number of consumers needed to obtain one complete set of comparisons.

---

*Many statisticians use the letter m in place of n when speaking of this problem. However the letter n is more in line with our regular notation.

EXAMPLE   Four brands can be compared using six consumers as follows:
15.6

	brand			
consumer	A	B	C	D
1	*	*		
2	*		*	
3	*			*
4		*	*	
5		*		*
6			*	*

Here an * indicates that the consumer will be asked to rate the indicated brand and compare it with the other starred brand. If $6r$ consumers are available, the same setup is repeated r times.

The analysis of data from such a paired comparison experiment is extremely simple. In each row we mark 1 for the preferred brand, leaving the other positions blank. This procedure corresponds to using "ranks" 0 and 1, rather than 1 and 2, when ranking the brands in each block. The effect is the same. Let R_i be the sum of the entries in the ith column. For a paired comparison experiment, R_i is simply the number of times brand i has been preferred in the various comparisons in which it has been used. In order to test the null hypothesis that consumers taken as a whole do not prefer any one brand over any other, we compute

$$Q = \frac{4}{rk} \{ [R_1 - \tfrac{1}{2}r(k-1)]^2 + \cdots + [R_k - \tfrac{1}{2}r(k-1)]^2 \}$$

$$= \frac{4}{rk} [R_1^2 + \cdots + R_k^2] - r(k-1)^2,$$

where r equals the number of complete comparison sets, a complete set requiring $k(k-1)/2$ consumers. The hypothesis is rejected if the observed value of Q surpasses the tabulated critical value of chi-square with $k-1$ degrees of freedom.

Multiple comparisons are performed by comparing rank sum differences $R_i - R_j$ with $\pm z\sigma$ where $\sigma = \sqrt{rk/2}$.

EXAMPLE   Forty consumers participated in a paired comparison experiment
15.7   involving the comparison of five brands of instant coffee. Here are their preferences:

$$R_1 = 7, \ R_2 = 14, \ R_3 = 11, \ R_4 = 3, \ R_5 = 5.$$

Thus brand 1 was preferred by 7 of the consumers who compared brand 1 with some other brand, brand 2 was preferred by 14 consumers, and so on. Can we conclude from these data that consumers have definite preferences among the various brands of instant coffee or are they simply making random choices?

Since we are comparing 5 brands, we have $k = 5$. According to Table 15.4 a complete set of comparisons requires 10 consumers. Since 40 consumers participated in the experiment we have $r = 4$.

Then

$$Q = \frac{4}{4 \times 5} (7^2 + 14^2 + 11^2 + 3^2 + 5^2) - 4 \times 4^2 = 16.$$

With 4 degrees of freedom, the result is highly significant. Apparently consumers are not making random guesses. We may then ask which brands are preferred to which. The multiple comparison method provides an answer. Using $\alpha = .20$ for a change, we find $\alpha/k(k-1) = \frac{.20}{20} = .01$ and therefore $z = 2.326$. Since σ equals $\sqrt{4 \times 5/2} = 3.162$, we find $z\sigma = 7.35$. Looking at differences $R_i - R_j$ we conclude that consumers prefer brand 2 to brands 4 and 5, and brand 3 to brand 4. However, when it comes to other pairs of brands, consumers do not have any common preferences.

PROBLEMS

1 Twenty test subjects each using one of four reducing diets lost the following number of pounds over a period of four weeks:

diet

A	B	C	D
2	9	4	18
7	10	17	14
22	8	0	24
18	20	6	13
15		11	12
			22

Is there any difference in the effectiveness of the four diets to produce loss of weight?

2 Six movie critics were asked to arrange four pictures A, B, C, and D in order of preference. The results were as follows:

critic

1	2	3	4	5	6
B	B	A	B	A	B
A	C	B	C	B	C
C	A	C	D	D	A
D	D	D	A	C	D

(For example, critic 1 preferred picture B over A over C over D.) Set up an appropriate hypothesis and test it.

3 In a paired comparison experiment, 30 consumers rated four kinds of soft drinks as follows:
12 preferred brand 1; 7, brand 2; 3, brand 3; and 8, brand 4.
Is there any kind of consensus among the consumers?

4 Plan and carry out a paired comparison experiment. Analyze your data as completely as possible.

5 A chess tournament is planned in which each of ten players plays twice against every other player.
   a. How many games does each player play?
   b. How many games will be played in all?

6 The experiment for comparing three methods of teaching algebra (Table 15.1a) could have been designed differently. If sufficient information about the pupils is available, we can divide 21 pupils into 7 groups of 3 in such a way that the 3 students in each group are as similar as possible with respect to factors thought to have a bearing on a pupil's learning ability for algebra. In each group, one student is randomly assigned to method 1, another to method 2, and a third to method 3. Suppose the examination scores for this experiment are as follows:

group	method 1	method 2	method 3
1	69	68	79
2	75	85	92
3	57	61	78
4	72	78	91
5	80	89	98
6	66	70	69
7	74	73	83

What are your conclusions? Perform as detailed an analysis as possible.

7 Suppose you want to compare the performance of six different suntan lotions using the services of the eight young ladies in Chapter 13. Design an appropriate experiment and indicate how you would analyze the resulting data.

8 Suppose that in Table 15.3a brands A and C are radial tires while brands B, D, and E are convential tires. Use the appropriate two-sample procedure to test the hypothesis that there is no difference in the stopping distances of the two types of tires.

9 Apply the method of multiple comparisons to the data in Table 15.1a.

10 Apply the method of multiple comparisons to the data in Table 15.3a using $\alpha = .10$.

11 Find the number of consumers needed to perform a paired comparison experiment involving
   a. 11 treatments,
   b. 12 treatments.

16

Rank Correlation

120   We return to the problem of investigating whether or not there exists a relationship between two characteristics. Is there a relationship between high school grades and college grades, for example? Presumably students who obtain high grades in high school also do well in college, but of course there are exceptions.

When we discussed the problem of relationship or association in Chapter 11, we merely assumed that each person or item belonging to our population could be classified according to two characteristics, each of which had a finite number of sub-classifications. We then used a chi-square test to test the hypothesis that no relationship existed. Our present example can be treated by these methods. For example, we can classify each student as having high, average, or low high school grades, and then use a similar classification for his college grades. But usually we have more precise information, like a student's grade point average, which is measured on a continuous scale. We should like to make use of this precise information, and in addition, we should like to measure strength of association, not merely test its absence.

We start with a relatively simple example, sort of half-way between a problem involving mere classifications and a problem involving continuous measurements. Suppose the eight young ladies who some time ago volunteered their services when we wanted to compare two suntan lotions are now participating in a contest for the title of Miss Suntan, sponsored by the manufacturers of the two suntan lotions. The president of each company is to rank all eight contestants, and the prize is to go to the one who obtains the highest combined score. Here are the results of the two rankings:

contestant	A	B	C	D	E	F	G	H
first ranking	8	4	6	7	2	3	1	5
second ranking	7	3	8	6	1	4	2	5
sum of ranks	15	7	14	13	3	7	3	10

The prize goes to Miss A, but Misses C and D provide rather close competition. The other contestants are clearly out of the running.

Going beyond the Miss Suntan contest, data of this type raise some interesting statistical questions. In particular, we may ask to what extent the two rankings agree. In Chapter 15 we encountered a somewhat similar situation. When discussing the problem of n rankings, we assumed that k objects had been ranked on n occasions. Then we wanted to find out whether or not there was any evidence of agreement among the n rankings. Now we have only two rankings, but we want to find an actual measure of the strength of agreement or disagreement between the two rankings.

Clearly our two rankings neither show complete agreement nor complete disagreement. Complete agreement means identical rankings:

$$1 \quad 2 \quad 3 \quad 4 \quad 5 \quad 6 \quad 7 \quad 8$$
$$1 \quad 2 \quad 3 \quad 4 \quad 5 \quad 6 \quad 7 \quad 8$$

Complete disagreement means reversed rankings:

$$1 \quad 2 \quad 3 \quad 4 \quad 5 \quad 6 \quad 7 \quad 8$$
$$8 \quad 7 \quad 6 \quad 5 \quad 4 \quad 3 \quad 2 \quad 1$$

Most observed rankings fall somewhere between these two extremes. Our example represents one of these in-between situations. But looking at the two rankings, we have the impression that they tend more towards perfect agreement than complete disagreement. This impression is strengthened when we re-arrange our data in such a way that one of the rankings, say the first, is in natural order:

$$1 \quad 2 \quad 3 \quad 4 \quad 5 \quad 6 \quad 7 \quad 8$$
$$2 \quad 1 \quad 4 \quad 3 \quad 5 \quad 8 \quad 6 \quad 7$$

The general trend of the second ranking is very similar to that of the first ranking. Is there any way in which we can express this intuitive feeling in a more precise fashion?

Statisticians have invented various types of *correlation coefficients* to measure strength of relationship. In our case since we are concerned with rankings, we speak of a *rank correlation coefficient*.

The Kendall Rank Correlation Coefficient

121 Rank correlation coefficients can be constructed in various ways. One of the simplest is the following. Suppose we count how often the two rankings move in the same direction and how often they move in opposite directions. We can accomplish this very simply. Let us look at our original data. Both judges ranked Miss A above Miss B, so we mark down +1 for the pair AB. The picture is different for the pair AC. The first judge preferred Miss A to Miss

C, but the second judge preferred Miss C to Miss A. We mark down −1 for this pair of contestants. We continue in this way for all 28 possible pairs of contestants, writing +1 if the two judges agree on the relative ranking of the pair of contestants, and writing −1 if they disagree. As a last step we find the sum S of all +1's and −1's. In our example there are 24 +1's and 4 −1's, so that $S = 24 - 4 = 20$ (see Problem 1).

This method of computing the number of agreements and disagreements among our two rankings is extremely simple. But it does take time, and we must be careful not to skip any pair. There is actually a much faster and safer way of computing the quantity S. We get the same value whether we compute S from our original table or from the rearranged table where the first ranking is in natural order. In the rearranged table, ranks in the top ranking always increase as we go from left to right. As a consequence, we need only mark +1 or −1 for a given pair of contestants depending on whether there is an increase or decrease in the second ranking as we go from the contestant on the left to the contestant further to the right.

The result $S = +20$ characterizes the degree of relationship between our two rankings. But how close is our result to perfect agreement or perfect disagreement? We can answer this question by computing S for the rankings representing these two situations. In the first case every pair contributes +1 for a total score of +28; in the second case every pair contributes −1 for a total score of −28. Thus $S = +28$ represents perfect agreement; $S = -28$ represents complete disagreement.

We can exhibit our result graphically. On a scale going from −28 to +28, our particular two rankings correspond to a value of +20:

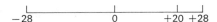

Our example shows the disadvantage of S as a measure of strength of relationship between the two rankings. We gain an impression of the strength of relationship only by considering S relative to its two possible extremes. This deficiency is easily remedied. All we have to do is divide S by its maximum value, that is, consider the new quantity $t = S/(\max S)$ in place of S. Then t is measured on a scale going from −1 to +1. For our example, $t = \frac{20}{28} = .71$, indicating clearly its position relative to the two extremes −1 and +1.

The quantity t is known as *Kendall's rank correlation coefficient*. A value of t near +1 implies close agreement among the rankings. A value of t near −1 implies almost diametrically opposite rankings. A value of t in the neighborhood of 0 indicates neither agreement nor disagreement. In this last case we may say that the two rankings are unrelated. Neither ranking can be used to throw light on the other.

122 We now want to find a formula for the quantity t in the general case where we rank n objects, rather than 8. Since by definition, $t = S/(\max S)$, we have to find $\max S$ in terms of n. Now S takes on its maximum value when every pair contributes $+1$. There are $(n - 1) + (n - 2) + \cdots + 2 + 1 = n(n - 1)/2$ pairs, so that

$$t = \frac{S}{\max S} = \frac{2S}{n(n - 1)}.$$

We can now answer the question that we raised in the beginning of this chapter. How can we measure the degree of relationship exhibited by two sets of observations when the observations are on a continuous scale? One useful way is to convert the two sets of observations to ranks and then compute the rank correlation coefficient t.

EXAMPLE 16.1

The final examination scores of 10 students in two subjects, psychology and statistics, are as follows:

TABLE 16.2

psychology	92	65	82	76	84	77	69	75	97	72
statistics	85	72	91	70	79	80	60	73	75	82

How closely do psychology and statistics grades agree? Our first step is to replace both sets of grades by their respective ranks. We then find the following two rankings:

TABLE 16.3

psychology	9	1	7	5	8	6	2	4	10	3
statistics	9	3	10	2	6	7	1	4	5	8

Now we rearrange the 10 rank pairs in such a way that the psychology ranking increases from 1 to 10:

TABLE 16.4

psychology	1	2	3	4	5	6	7	8	9	10
statistics	3	1	8	4	2	7	10	6	9	5

Mere inspection shows that the agreement in the two rankings is far from perfect. But let us compute the quantity S. The first statistics rank is 3. Further to the right there are 7 increases and 2 decreases for a total of $7 - 2 = 5$. The second statistics rank is 1, and further to the right there are 8 increases and no decreases for a total of $8 - 0 = 8$. Continuing in this way we find

$$S = 5 + 8 - 3 + 4 + 5 + 0 - 3 + 0 - 1 = 15.$$

And as a last step,

$$t = \frac{2S}{n(n - 1)} = \frac{2 \times 15}{10 \times 9} = \frac{1}{3}.$$

A rank correlation of $\frac{1}{3}$ indicates some agreement among the two sets of grades. On the whole students who get high grades in psychology also get high grades in statistics and vice versa. But there are some marked exceptions. For example, one student received 97 in psychology but only 75 in statistics.

123 In doing Example 16.1 our first step consisted of converting the data in Table 16.2 to the ranks in Table 16.3 and then rearranging the psychology ranks from smallest to largest. We then computed S from the rearranged ranks in Table 16.4. Actually the conversion of the original data to ranks is not needed for the computation of S. This step was introduced only to bring out the rank nature of the correlation coefficient t. It is sufficient to rearrange the original data in Table 16.2 according to increasing psychology grades as follows:

TABLE 16.5

psychology	65	69	72	75	76	77	82	84	92	97
statistics	72	60	82	73	70	80	91	79	85	75

We can then compute S exactly as before by noting the number of increases and decreases when comparing each statistics grade in Table 16.5 with all other statistics grades further to the right.

124 ***Tied Observations*** Our method for computing S has assumed that no two or more observations in either sequence are equal. If ties occur in one or both sequences, when computing S we count 0 for a pair in which both numbers in a sequence are equal.

EXAMPLE 16.6

Find the Kendall rank correlation coefficient for the following two sets of grades:

student	A	B	C	D	E
grade in subject 1	65	71	73	80	86
grade in subject 2	70	75	70	82	85

Since students A and C obtained identical grades in subject 2, the pair (A, C) contributes 0 when computing S. Thus

$$S = 3 + 1 + 2 + 1 = 7$$

and

$$t = \frac{2S}{5 \times 4} = .7.$$

A Test of Independence

125 Example 16.1 suggests another problem that requires our attention. Presumably the ten students whose grades in psychology and statistics we have just investigated constitute only a small fraction of all students who take these two courses. What can we say about the relationship between psychology and statistics grades in general? In particular, is it possibly true that grades in these two subjects are unrelated and that the observed value $t = \frac{1}{3}$ has arisen due to chance? This calls for a test of independence. Only if the observed value of t, and consequently of S, differs significantly from 0 can we claim that psychology and statistics grades are not independent. As on several earlier occasions, an appropriate z-statistic gives the answer:

$$z = \frac{6S}{\sqrt{2n(n-1)(2n+5)}} = \frac{6 \times 15}{\sqrt{2 \times 10 \times 9 \times 25}} = 1.34.$$

The descriptive level associated with this z-value (using a two-sided test) is $2 \times .090 = .180$, so that we cannot rule out the possibility that for the student population as a whole psychology and statistics grades are unrelated.

The normal approximation will usually be adequate for this test, as long as n is at least 8. For smaller samples, a table of the exact S-distribution should be consulted.

A Test of Randomness against a Monotone Trend

126   Implicit in all procedures discussed in earlier chapters is the assumption that observations in a sample come from one and the same population and that the order in which the observations are observed can be ignored. There are clearly experiments for which this assumption is invalid. If we measure the weight of a test animal at weekly intervals, we should expect to observe random fluctuations from one week to another, as well as a general upward trend during the growth period of the animal. There are practical situations when it is important to find out whether or not a given set of observations exhibits such a monotone trend. The previous test of independence can be used to test the hypothesis that observations in a sample are purely random against the alternative of a monotone trend.

Let $x_1, x_2, \ldots, x_n$ be the set of observations (obtained in this order) suspected of following either an upward or downward trend. Let r_i be the rank of x_i. We then compute the quantity S for the following two rankings:

$$\begin{array}{cccc} 1 & 2 & \cdots & n \\ r_1 & r_2 & \cdots & r_n \end{array}$$

If the observations $x_1, x_2, \ldots, x_n$ do not exhibit a downward or upward trend, the two rankings are unrelated, resulting in a value of S in the neighborhood of zero. On the other hand, a large positive value of S is indicative of an upward trend; a large negative value of S, of a downward trend. The test procedure discussed in the preceding section then becomes a test of randomness against a monotone trend.

EXAMPLE
16.7
Mr. A attended a course advertised to increase one's reading speed. At the end of each of 12 sessions Mr. A's reading speed in words per minute was recorded as follows (reading left to right):

$$495 \quad 525 \quad 520 \quad 490 \quad 555 \quad 530$$
$$475 \quad 510 \quad 515 \quad 545 \quad 540 \quad 550$$

Would you say that the course achieved its advertised purpose? In order to answer this question, we test the hypothesis of randomness against the one-sided alternative of an upward trend. Rejection of this hypothesis implies that Mr. A's reading speed has gradually increased over the period of instruction.

We first convert the observations to ranks:

$$3 \quad 7 \quad 6 \quad 2 \quad 12 \quad 8 \quad 1 \quad 4 \quad 5 \quad 10 \quad 9 \quad 11$$

The quantity S is now easily found:

$$
\begin{aligned}
S = {} & (9-2) + (5-5) + (5-4) + (7-1) \\
& + (0-7) + (3-3) + (5-0) + (4-0) \\
& + (3-0) + (1-1) + (1-0) \\
= {} & 20
\end{aligned}
$$

(Again this computation could have been carried out without converting the original observations to ranks. For example, the difference $(9-2)$ simply says that 9 of the observations to the right of 495 are greater than 495, while 2 are less than 495.) It follows that

$$z = \frac{6 \times 20}{\sqrt{2 \times 12 \times 11 \times 29}} = 1.37.$$

The descriptive level of our test is given by the area under the normal curve to the right of $z = 1.37$. This area is .085. Our evidence for believing that the course has increased Mr. A's reading speed is not very strong.

PROBLEMS
1 Check the value of S for the Miss Suntan contest, using both methods of computation suggested in the text.

2 Find the rank correlation coefficient for the two rankings of critics 3 and 4 in Problem 2 of Chapter 15.

3 Two judges at a figure skating contest assigned place numbers as follows to ten contestants:

judge A	2	5	6	4	1	7	9	10	3	8
judge B	1	4	5	6	2	7	10	8	3	9

a. Find the rank correlation coefficient.

b. Test for independence. What conclusion can you draw from the test result?

4 Find the rank correlation coefficient for the test scores for methods 1 and 2 in Problem 6 of Chapter 15. Do the same for methods 1 and 3. Would you say that the experimenter has been successful in arranging the 21 pupils in homogeneous groups?

5 Twelve students applying for admission to a graduate school showed the following scores on the verbal and quantitative parts of the Graduate Record Examination (the first number in each pair represents the verbal score; the second, the quantitative score):

(560, 620), (605, 490), (585, 552), (678, 632)
(621, 512), (482, 780), (615, 665), (530, 545)
(633, 545), (618, 730), (575, 588), (672, 642)

a. Find the rank correlation coefficient.

b. Test for independence.

6 For a sample of size 15 you have found $S = -32$. Using a significance level of .05, test the hypothesis of randomness against the alternative of

a. a monotone trend,

b. a downward trend.

7 You have made eight successive determinations of the salinity of a body of water:

2.33   2.24   2.29   2.20   2.18   2.25   2.19   2.15

Would you say that the degree of salinity is decreasing over time?

8 In Problem 4 of Chapter 12 you were asked to determine the heights and weights of 12 randomly selected students. Test the hypothesis that height and weight are independent in the student population.

9 In connection with the draft lottery of December 1, 1969 it was generally stated that men with draft numbers 1–183 would very likely be called, while those with draft numbers 184–366 would much less likely be called. The following table gives the number of draft dates in the first category for each month of the year:

Jan	Feb	Mar	Apr	May	Jun	Jul	Aug	Sep	Oct	Nov	Dec
12	12	10	11	14	14	14	19	17	13	21	26

Find the Kendall rank correlation coefficient between the number of draft dates in a month and the rank of the month in the year. What possibility does the result suggest?

10 Two contest judges ranked five contestants as follows:

contestant	A	B	C	D	E
judge 1	1	2.5	2.5	4	5
judge 2	1	2.5	2.5	4	5

Show that $t = .9$. Note that even though the two judges used identical rankings, the value of t is less than $+1$. This is due to the fact that the two judges could not distinguish between contestants B and C.

17

Samples from Normal Populations

127 The nonparametric procedures discussed in Chapters 12–16 require only the weakest of assumptions about the populations being sampled. If more detailed information is available, more specific procedures for solving the problems of Chapters 12–16 can be developed. In the remaining chapters of the book we assume that we have samples from normal populations. Now we reformulate our earlier problems in terms of normal distributions and make some general remarks about nonparametric versus normal theory methods.

The mean μ of a normal distribution is also its median. As a consequence, the methods developed in Chapter 12 remain applicable by the simple expedient of replacing η by μ. We might then ask why we should want to develop further procedures. As we have already pointed out in Chapter 12, the answer is that by restricting ourselves to normal or near-normal populations we can hope to develop more powerful tools than the nonparametric methods that were developed to serve under much more general conditions. By more powerful we mean point estimates that on the average are closer to the true parameter value; confidence intervals that for the same confidence coefficient tend to be shorter; tests of hypotheses that for the same significance level have smaller type 2 error probabilities. On the other hand, when dealing with nonnormal data, use of normal theory procedures not only may change the significance level of a test or the confidence coefficient of a confidence interval, but also may actually make less effective use of the available sample information than appropriate nonparametric methods.

Actually, there may be reasons for preferring nonparametric procedures to normal theory methods even when the normality

163

assumption seems justified. Due to occurrences such as mis-reading, careless experimentation, inadequate controls, and the like, it may happen that one or two observations differ rather markedly from the rest of the observations. Such observations are often referred to as *outliers*. Outliers cause complications in statistical analysis since, in practice, we can rarely completely rule out the possibility that a suspected outlier has arisen in a natural way. There is no fully satisfactory method for dealing with such observations. However, a good way to guard against possible misleading effects is to use a method of statistical analysis that is relatively unaffected by outliers. In general, nonparametric procedures are considerably less affected by outliers than are normal theory procedures.

128 In practice the mean μ of a normal population is of greater interest than the standard deviation σ. For this reason, in Chapter 18, we spend considerable time on statistical methods involving μ. Problems involving the estimation of and tests about the standard deviation σ are considered much more briefly.

Chapters 13 and 14 were concerned with the comparison of two populations. We now assume that both populations are normal, the first, or x-population, with mean μ_x and standard deviation σ_x; the second, or y-population, with mean μ_y and standard deviation σ_y. Of greatest practical interest are comparisons between μ_x and μ_y. The tests of Chapters 13 and 14 can be considered tests of the hypothesis $\mu_x = \mu_y$, while the corresponding confidence procedures provide confidence intervals for the difference $\Delta = \mu_x - \mu_y$. In Chapter 19 we discuss methods of analysis that are based specifically on the assumption of normality. We also discuss a test of the hypothesis $\sigma_x = \sigma_y$.

The methods of Chapter 15 become tests of the hypothesis $\mu_1 = \cdots = \mu_k$ when applied to samples from k normal populations with means $\mu_1, \ldots, \mu_k$. The corresponding normal theory analysis to be discussed in Chapter 20 is based on a technique known as the *analysis of variance*. We also discuss simultaneous confidence intervals.

Finally, the problems of Chapter 16 lead to the consideration of regression and normal correlation in Chapter 21.

18

One-Sample Problems for Normal Populations

Point Estimates for the Parameters of a Normal Distribution

129 We start our discussion of normal theory methods with point estimates of the mean and standard deviation. Let $x_1, \ldots, x_n$ represent a random sample from a normally distributed population with mean μ and standard deviation σ. We shall use the following point estimates of μ and σ:

18.1
$$\bar{x} = \frac{x_1 + \cdots + x_n}{n} = \frac{1}{n} \sum_j x_j$$

and

18.2
$$s = \sqrt{\frac{1}{n-1} \sum_j (x_j - \bar{x})^2}.^*$$

These estimates are called the *sample mean* and *sample standard deviation* respectively.

We have seen that μ is the center of the normal distribution. The sample mean $\bar{x}$ is the arithmetic mean of the observations and as such is a measure of centrality for the sample observations.

The standard deviation σ is a measure of variability. Normal distributions with large standard deviations are low and wide, indicating a population that is spread out. Normal distributions with small standard deviations are high and narrow, indicating a population that is closely concentrated about its mean. The sample standard deviation gives similar information about the

---

*Students who are unfamiliar with summation notation should read Section 29.

sample observations. A large value of s indicates that at least some of the observations are a considerable distance away from the mean $\bar{x}$. A small value of s indicates that all observations are close to the sample mean. Formula 18.2 shows that in order to compute s we first have to find the *sample variance*

18.3a
$$s^2 = \frac{1}{n-1} \sum_j (x_j - \bar{x})^2,$$

which is an estimate of the population variance σ^2.

130   ***Computation of s^2***   Formula 18.3a for s^2 is primarily useful for purposes of definition. When we want to compute s^2 from given sample observations $x_1, \ldots, x_n$, an equivalent formula is often preferable:

18.3b
$$s^2 = \frac{1}{n-1} \left[\sum_j x_j^2 - \frac{1}{n} \left(\sum_j x_j \right)^2 \right].$$

The equivalence of (18.3a) and (18.3b) is seen as follows:

18.4
$$
\begin{aligned}
\sum_j (x_j - \bar{x})^2 &= \sum_j (x_j^2 - 2\bar{x}x_j + \bar{x}^2) \\
&= \sum_j x_j^2 - 2\bar{x} \sum_j x_j + n\bar{x}^2 \\
&= \sum_j x_j^2 - \frac{2}{n} \left(\sum_j x_j \right)^2 + \frac{1}{n} \left(\sum_j x_j \right)^2 \\
&= \sum_j x_j^2 - \frac{1}{n} \left(\sum_j x_j \right)^2,
\end{aligned}
$$

and (18.3b) follows.

EXAMPLE 18.5a   Find the sample mean, variance, and standard deviation for the following five observations: 75, 82, 68, 87, 79.
We have
$$\sum_j x_j = 75 + 82 + 68 + 87 + 79 = 391$$
and
$$\sum_j x_j^2 = 75^2 + 82^2 + 68^2 + 87^2 + 79^2 = 30783$$
so that
$$\bar{x} = \frac{391}{5} = 78.2,$$
$$\sum_j (x_j - \bar{x})^2 = 30783 - \frac{391^2}{5} = 206.8,$$
$$s^2 = \frac{206.8}{4} = 51.7 \text{ and } s = \sqrt{51.7} = 7.2.$$

131   Actually, computations can often be simplified by subtracting a suitable constant from all observations. Before we see how, we have to investigate how the subtraction of the same constant c from all observations affects the sample variance. Let

18.6
$$y_1 = x_1 - c, \ldots, y_n = x_n - c.$$

Clearly the subtraction of c from all x-observations subtracts c from the arithmetic mean,

18.7
$$\bar{y} = \bar{x} - c.$$

If we use s_y^2 to denote the sample variance of the y's, we have, using (18.6) and (18.7),

$$
\begin{aligned}
s_y^2 &= \frac{1}{n-1} \sum_i (y_i - \bar{y})^2 \\
&= \frac{1}{n-1} \sum_i [(x_i - c) - (\bar{x} - c)]^2 \\
&= \frac{1}{n-1} \sum_i [x_i - c - \bar{x} + c]^2 \\
&= \frac{1}{n-1} \sum_i (x_i - \bar{x})^2 = s_x^2.
\end{aligned}
$$

The y's have exactly the same sample variance as the x's.

132   We can use this fact when computing the sample variance of a set of observations $x_1, \ldots, x_n$ as follows. We guess roughly the mean of the observations and use our guess as the constant c. The sample variance of the y's resulting from the subtraction of c equals the sample variance for the original x's. We can usually choose c in such a way that the y's are much more manageable numbers than the x's, thus simplifying the computational effort.

EXAMPLE
18.5b

For the data of Example 18.5a we may choose $c = 80$. The y's then become

$$-5, 2, -12, 7, -1.$$

We easily find $\sum_i y_i = -9$ and $\sum_i y_i^2 = 223$. Thus $\bar{y} = -\frac{9}{5} = -1.8$ and by (18.7),

$$\bar{x} = \bar{y} + c = -1.8 + 80 = 78.2.$$

Further, applying (18.3b) to the y's,

$$s_x^2 = s_y^2 = \frac{\sum_i y_i^2 - (\sum_i y_i)^2/n}{n-1} = \frac{223 - 9^2/5}{4} = 51.7.$$

While the computations for the y's are considerably faster and simpler, the method has one disadvantage. To get the y's we have to perform n subtractions, each representing a potential source of error.

133 **The Sampling Distributions of $\bar{x}$ and s^2** In Chapter 7 when we discussed estimation of the success probability p for binomial trials, we pointed out that point estimates by themselves are of little use unless they are accompanied by an indication of the amount of fluctuation that can be expected from sample to sample. For $\bar{x}$ and s^2 such information is obtained from the following theorem proved in courses in mathematical statistics:

THEOREM *In random samples from normal populations with mean μ and*
18.8 *standard deviation σ, the sample mean $\bar{x}$ is normally distributed with mean μ and standard deviation $\sigma/\sqrt{n}$. Further, the quantity $(n-1)s^2/\sigma^2 = \sum_j (x_j - \bar{x})^2/\sigma^2$ has a chi-square distribution with $n-1$ degrees of freedom.*

According to the first part of the theorem, the quantity

18.9
$$z = \frac{\bar{x} - \mu}{\sigma/\sqrt{n}}$$

has a standard normal distribution. As a consequence the following statement holds with probability γ:

18.10a
$$-z_\gamma \leq \frac{\bar{x} - \mu}{\sigma/\sqrt{n}} \leq z_\gamma,$$

where z_γ is read from Table C corresponding to γ. Inequality (18.10a) can be written in the form

18.10b
$$-z_\gamma \sigma/\sqrt{n} \leq \bar{x} - \mu \leq z_\gamma \sigma/\sqrt{n}$$

or equivalently,

18.10c
$$|\bar{x} - \mu| \leq z_\gamma \sigma/\sqrt{n}.$$

The inequality in (18.10c) has the following meaning. In random samples of size n from a normal population with standard deviation σ, the point estimate $\bar{x}$ deviates from the population mean μ by no more than $d = z_\gamma \sigma/\sqrt{n}$, unless an event of probability $1 - \gamma$ occurs. By taking a sufficiently large number of observations, we can make d as small as we please. In particular, we can determine n such that d has a prescribed value d_0. This is achieved if $z_\gamma \sigma/\sqrt{n} = d_0$ or $n = (z_\gamma \sigma/d_0)^2$.

Confidence Intervals for the Mean of a Normal Distribution

134 **Population Standard Deviation Known** In order to find a confidence interval for the population mean μ, we write (18.10c) as

$$|\mu - \bar{x}| \leq z_\gamma \sigma/\sqrt{n},$$

or equivalently as

$$-z_\gamma \sigma/\sqrt{n} \leq \mu - \bar{x} \leq z_\gamma \sigma/\sqrt{n}.$$

Adding $\bar{x}$ to all three parts, we find the following confidence interval for μ,

18.11
$$\bar{x} - z_\gamma \sigma / \sqrt{n} \leq \mu \leq \bar{x} + z_\gamma \sigma / \sqrt{n}.$$

EXAMPLE 18.12a   Assume that the 5 observations in Example 18.5a have come from a normal population with unknown mean μ, but known standard deviation $\sigma = 10$. We want to find a confidence interval with confidence coefficient .95 for μ. We compute $z_\gamma \sigma / \sqrt{n} = 1.96 \times 10/\sqrt{5} = 8.8$ and (18.11) becomes

$$78.2 - 8.8 \leq \mu \leq 78.2 + 8.8$$

or

$$69.4 \leq \mu \leq 87.0.$$

135   We have remarked repeatedly that the length of a confidence interval is a good indicator of its usefulness. The length of the interval (18.11) is

$$L = (\bar{x} + z_\gamma \sigma / \sqrt{n}) - (\bar{x} - z_\gamma \sigma / \sqrt{n}) = 2z_\gamma \sigma / \sqrt{n}.$$

We can use this formula to determine the sample size required to produce a confidence interval of prescribed length L_0. Indeed we must have $2z_\gamma \sigma / \sqrt{n} = L_0$, so that

$$n = \left(\frac{2z_\gamma \sigma}{L_0} \right)^2.$$

As on earlier occasions, we observe that a reduction of the length of a confidence interval by a factor of two requires four times as many observations; by a factor of three, nine times as many observations.

EXAMPLE 18.12b   Under the assumptions of Example 18.12a, how many observations are required to produce a confidence interval of length 6 with $\gamma = .95$? Substituting in the formula for n we find $n = (2 \times 1.96 \times 10/6)^2 = 42.7$, which we round off to the next largest integer, 43.

136   ***Population Standard Deviation Unknown***   The problem of what to do when the population standard deviation is unknown or is inaccurately known, as happens often in practice, is easily solved when we are dealing with large samples. In that case the sample standard deviation s is a good point estimate of the population standard deviation σ, and replacement of σ by s has little influence on the confidence coefficient γ. Thus in large samples the following confidence interval is usually satisfactory:

18.13
$$\bar{x} - z_\gamma s / \sqrt{n} \leq \mu \leq \bar{x} + z_\gamma s / \sqrt{n}.$$

However when n is smaller than 30 or 40, this solution generally is unsatisfactory, since the true confidence coefficient associated with the interval in (18.13) is smaller than is implied by z_γ. For example, if $n = 6$ and we use $z_\gamma = 2$, corresponding to the value $\gamma = .955$ according to Table C, the true confidence coefficient associated with the interval in (18.13) is .90.

We can adjust for this deficiency by replacing the quantity z_γ by a somewhat larger quantity t_γ, which is read from a table of the so-called t-distribution with $n - 1$ degrees of freedom. Appropriate t-values are found in Table I. Inspection of Table I shows that if n is as large as 40, there is very little difference between t_γ and z_γ (which is tabulated in Table I under ∞).

More specifically, if in (18.9) we replace the population standard deviation σ by the sample estimate s, we obtain the quantity

18.14
$$t = \frac{\bar{x} - \mu}{s/\sqrt{n}},$$

which has the t-distribution tabulated in Table I with $n - 1$ degrees of freedom, the number of degrees of freedom associated with the estimate s (see Theorem 18.8). The confidence interval (18.11) becomes

18.15
$$\bar{x} - t_\gamma s/\sqrt{n} \leq \mu \leq \bar{x} + t_\gamma s/\sqrt{n}.$$

EXAMPLE
18.12c
If in Example 18.12a we assume that the population standard deviation is unknown, we obtain the following confidence interval:

$$78.2 - 2.776 \times 7.2/\sqrt{5} \leq \mu \leq 78.2 + 2.776 \times 7.2/\sqrt{5}$$

or

$$69.3 \leq \mu \leq 87.1.$$

Note that this confidence interval is slightly wider than the interval in Example 18.12a, even though the sample estimate s happens to be considerably smaller than the population standard deviation σ assumed in Example 18.12a. When the population standard deviation is unknown, we have to expect wider intervals than when the standard deviation is known.

Tests of Hypotheses about the Mean of a Normal Population

137
The hypothesis $\mu = \mu_0$ can be tested against the two-sided alternative $\mu \neq \mu_0$ at significance level $\alpha'' = 1 - \gamma$ by finding out whether or not the hypothetical value μ_0 is in the appropriate confidence interval. If σ is known, the interval in (18.11) is used; if σ is unknown, the interval in (18.15) is appropriate.

A direct test of the hypothesis $\mu = \mu_0$ proceeds as follows. We replace the parameter μ in (18.9) or (18.14) by the hypothetical value μ_0 and reject the hypothesis when the resulting z- or t-value is critical according to the z- or t-distribution. The confidence interval procedures of Sections 134 and 136 test the hypothesis $\mu = \mu_0$ against the two-sided alternative $\mu \neq \mu_0$. The direct test can be used against one-sided alternatives as well.

EXAMPLE
18.16a

We want to test the hypothesis $\mu = 70$ at significance level .05 using the assumption of Example 18.12a. We find

$$z = \frac{78.2 - 70}{10/\sqrt{5}} = 1.83.$$

The two-sided critical region consists of z-values less than -1.96 or greater than $+1.96$. Thus we accept the hypothesis $\mu = 70$ when testing against the alternative $\mu \neq 70$. Against the one-sided alternative $\mu > 70$, we reject the hypothesis $\mu = 70$ only for sufficiently large values of z, namely, $z > 1.645$. Our observed z-value indicates rejection under these circumstances.

EXAMPLE
18.16b

We consider the same problem as in Example 18.16a, now assuming that the population standard deviation is unknown. This time we find

$$t = \frac{78.2 - 70}{7.2/\sqrt{5}} = 2.55.$$

For the two-sided test the critical values are -2.776 and $+2.776$, indicating acceptance of the null hypothesis. The one-sided critical value is $+2.132$, indicating rejection.

EXAMPLE
18.17

When testing the hypothesis $\mu \geq \mu_0$ against the alternative $\mu < \mu_0$ using a random sample of size 9 from a normal population with unknown standard deviation, we reject the null hypothesis at significance level .01 provided $t = 3(\bar{x} - \mu_0)/s < -2.896$.

The Central Limit Theorem

138   The procedures discussed so far in this chapter are based on the theorem that in random samples from a normal population with mean μ and standard deviation σ, the sample mean $\bar{x}$ is normally distributed with mean μ and standard deviation $\sigma/\sqrt{n}$. One of the most remarkable and useful theorems of the theory of probability, the *Central Limit Theorem*, states that this result remains approximately true for random samples from (almost) any kind of population provided the number of observations n in the sample is sufficiently large. In large samples we do not even have to assume that the population standard deviation is known. Using the sample standard deviation s in place of the population standard deviation σ usually provides satisfactory results. It then follows that (18.13) provides a confidence interval for the mean μ of an arbitrary population, provided the number of observations is sufficiently large. Alternatively, the hypothesis $\mu = \mu_0$ can be tested by treating the statistic $\sqrt{n}(\bar{x} - \mu_0)/s$ as a standard normal variable.

Inference Procedures for the Standard Deviation

139   **A Confidence Interval for σ**   According to Theorem 18.8, in random samples of n observations from a normal distribution with standard deviation σ, the quantity

$$\frac{(n-1)s^2}{\sigma^2} = \frac{1}{\sigma^2}\sum_j(x_j - \bar{x})^2$$

has a chi-square distribution with $n-1$ degrees of freedom. This result can be used to find a confidence interval for σ having confidence coefficient γ as follows. Let $\alpha = 1 - \gamma$. From a table of the chi-square distribution we find two values V_1 and V_2 such that the area to the left of V_1 is $\alpha/2$ and the area to the right of V_2

FIGURE 18.18

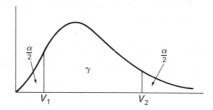

is also $\alpha/2$. Then with probability γ,

$$V_1 \leq \frac{\sum_j(x_j - \bar{x})^2}{\sigma^2} \leq V_2,$$

which can also be written as

$$\frac{1}{V_2} \leq \frac{\sigma^2}{\sum_j(x_j - \bar{x})^2} \leq \frac{1}{V_1}.$$

Multiplying all three parts by $\sum_j(x_j - \bar{x})^2$, we find the confidence interval for σ^2,

$$\frac{\sum_j(x_j - \bar{x})^2}{V_2} \leq \sigma^2 \leq \frac{\sum_j(x_j - \bar{x})^2}{V_1}.$$

But then

18.19

$$\sqrt{\frac{\sum_j(x_j - \bar{x})^2}{V_2}} \leq \sigma \leq \sqrt{\frac{\sum_j(x_j - \bar{x})^2}{V_1}}$$

is a confidence interval for σ.

EXAMPLE 18.20a   We want to find a confidence interval for σ using the data of Example 18.5a. Corresponding to the confidence coefficient .95 we find $V_1 = .484$ and $V_2 = 11.1$. Thus the lower confidence bound for σ^2 is $206.8/11.1 = 18.6$; the upper bound is $206.8/.484 = 427.3$. The corresponding bounds for σ are $\sqrt{18.6} = 4.3$ and $\sqrt{427.3} = 20.7$.

140 *Testing a Hypothesis about* σ The confidence interval in (18.19) provides a two-sided test of the hypothesis $\sigma = \sigma_0$. Alternatively this hypothesis can be tested by referring the statistic $\sum_j (x_j - \bar{x})^2/\sigma_0^2$ directly to a table of the chi-square distribution with $n - 1$ degrees of freedom.

EXAMPLE 18.20b We want to test the hypothesis $\sigma = 10$ using the data of Example 18.5a. The appropriate test statistic is $\sum_j (x_j - \bar{x})^2/\sigma_0^2 = 206.8/10^2 = 2.068$. Suppose we have decided on a significance level .05. When testing against the two-sided alternative $\sigma \neq 10$, we accept the null hypothesis, since the test statistic has a value between .484 and 11.1. This decision agrees of course with the confidence interval of Example 18.20a. As a second illustration consider the case where we should like to find out whether the true standard deviation is less than 10. We would then test the hypothesis $\sigma \geq 10$ against the alternative $\sigma < 10$. Using again a significance level .05, the test statistic would have to be smaller than .711 for rejection, so we accept the hypothesis. It would seem that the true standard deviation is at least 10.

In solving Examples 18.20a and 18.20b, we have assumed that the five observations represent a random sample from some normal distribution.

PROBLEMS

1 Find the sample mean for the data of
   a. Example 12.5a,
   b. Example 12.6,
   c. Problem 5, Chapter 12.
   In each case compare the mean with the median for the same set of data. How do you explain that in some cases the mean and median are close together, while in others they are not?

2 Find the sample standard deviation for the data of
   a. Example 12.5a,
   b. Example 12.6,
   c. Problem 5, Chapter 12.

3 To what extent do the test results in Examples 18.16a and 18.16b agree with the confidence intervals in Examples 18.12a and 18.12c?

4 Assume that the data in Problem 3 of Chapter 12 represent a random sample from a normal population with standard deviation 10.
   a. Find a confidence interval having confidence coefficient .98 for the population mean μ.
   b. Test the hypothesis $\mu = 75$ at significance level .05.
   c. How many observations are required to obtain a 99% confidence interval of length 5?

5 Solve parts (a) and (b) of the preceding problem without making any assumption about the population standard deviation.

6 Solve Problems 5 and 7 of Chapter 12 using a normal theory test. Is the normal theory test appropriate in each of the two cases?

7 Do Problem 6 of Chapter 12 assuming samples from normal populations.

8 For the tire data in Problem 5 of Chapter 12
   a. test the hypothesis that $\sigma = 2500$,
   b. find a confidence interval for σ.
   What assumption is implicit in your solution?

9 Each student selects a random sample of size 16 from Table L of normal random deviates and adds some arbitrary constant μ to each of the 16 observations. He then asks a second student to find a 90% confidence interval for μ. For the class as a whole, what proportion of the computed confidence intervals contain the true value μ added by the first student?

10 Each student takes a random sample of size 25 from Table L of normal random deviates. He then performs the following two steps in the indicated order: (i) multiply each observation by the same positive constant σ; (ii) add the same constant μ to each product. He then asks a second student to find 95% confidence intervals for μ and σ. For the class as a whole, what proportion of the computed confidence intervals contain the true values of μ and σ?

*11 Let $X_1, \ldots, X_n$ be n independent random variables, each having mean μ and variance σ^2. Let $\overline{X} = (X_1 + \cdots + X_n)/n$.
   a. Show that the mean of $\overline{X}$ is μ. (*Hint:* Use Theorems 4.1 and 4.2.)
   b. Show that the variance of $\overline{X}$ is σ^2/n. (*Hint:* Use Theorem 4.3 and Problem 10 of Chapter 4.)

19

Two-Sample Problems for Normal Populations

141 In this chapter we develop normal theory solutions for the type of problems discussed in Chapters 13 and 14. As we pointed out in Chapter 17, in situations when samples from two normal populations are available, the investigator is usually most interested in the difference of means $\Delta = \mu_x - \mu_y$. But we shall also treat, though much more briefly, the problem of comparing the two standard deviations σ_x and σ_y.

The following theorem, proved in courses on mathematical statistics, is needed for our discussion:

THEOREM   *If u and v are two (jointly) normally distributed variables with*
19.1   *means μ_u and μ_v and standard deviations σ_u and σ_v, then the variable $w = u - v$ is again normally distributed with mean $\mu_w = \mu_u - \mu_v$. If in addition u and v are independent, then the standard deviation of w equals $\sigma_w = \sqrt{\sigma_u^2 + \sigma_v^2}$.*
(If u and v are not independent, the standard deviation of w is a more complicated expression.)

Paired Observations

142 We first consider the situation in Chapter 13, when x- and y-observations occur in natural pairs: $x_1, y_1; x_2, y_2; \ldots; x_n, y_n$. As before, we consider only the differences $d_1 = x_1 - y_1$, $d_2 = x_2 - y_2, \ldots, d_n = x_n - y_n$. The theorem that we have just mentioned implies that the set of differences $d_1, d_2, \ldots, d_n$ represents a random sample from a normal population with mean Δ. (It is usually appropriate to consider the standard deviation σ_d unknown.) Thus we have reduced the two-sample problem to a one-sample problem, which can be solved by the methods of Chapter 18.

175

By (18.1) a point estimate of Δ is given by

$$\overline{d} = \frac{1}{n}\sum_j d_j.$$

By (18.15) a confidence interval for Δ is given by

$$\overline{d} - t_\gamma s_d/\sqrt{n} \leq \Delta \leq \overline{d} + t_\gamma s_d/\sqrt{n},$$

where, by (18.2),

$$s_d = \sqrt{\frac{1}{n-1}\sum_j(d_j - \overline{d})^2}.$$

Finally, by (18.14) the hypothesis $\Delta = \Delta_0$ is tested by means of the t-statistic

$$t = \frac{(\overline{d} - \Delta_0)}{s_d/\sqrt{n}}.$$

EXAMPLE
19.2

In the suntan lotion experiment of Chapter 13 we observed the following differences: 5, 3, −1, 14, 7, −4, 13, 10. It is usually not unreasonable to assume that measurements of this type are approximately normally distributed. We shall therefore reanalyze the given data using the normal theory approach we have just developed.

We find $\overline{d} = \frac{47}{8} = 5.875$. Since the accuracy of the original observations is to the nearest integer, it is inappropriate to keep more than one decimal place when estimating Δ on the basis of only eight observations. Thus our estimate is 5.9.

For a 95% confidence interval we have $t_\gamma s_d/\sqrt{n} = 2.365 \times 6.424/\sqrt{8} = 5.4$ so that

$$5.9 - 5.4 \leq \Delta \leq 5.9 + 5.4$$

or

$$0.5 \leq \Delta \leq 11.3.$$

Finally, the test of the hypothesis $\Delta = 0$ is based on the statistic

$$t = \frac{\overline{d}}{s_d/\sqrt{n}} = 2.59,$$

indicating rejection of the hypothesis at significance level .05, a result that is already implied by the confidence interval.

Independent Samples

143 **_Known Population Standard Deviations_**   When developing normal alternatives for the methods of Chapter 14, we assume that the observations $x_1, \ldots, x_m$ represent a random sample from a normal population with mean μ_x and standard deviation σ_x, and the observations $y_1, \ldots, y_n$, an independent random sample from a normal population with mean μ_y and standard deviation σ_y.

An obvious point estimate of $\Delta = \mu_x - \mu_y$ is the difference of sample means $\bar{x} - \bar{y}$. In order to find a confidence interval for Δ, we set $u = \bar{x}$ and $v = \bar{y}$ in Theorem 19.1. It then follows that $\bar{x} - \bar{y}$ is normally distributed with mean $\mu_x - \mu_y = \Delta$ and standard deviation $\sqrt{\sigma_x^2/m + \sigma_y^2/n}$. In particular, if $\sigma_x = \sigma_y = \sigma$, the standard deviation reduces to

$$\sigma \sqrt{\frac{1}{m} + \frac{1}{n}} = \sigma \sqrt{\frac{m+n}{mn}}.$$

Thus if σ_x and σ_y are known, a confidence interval for Δ is given by

19.3a
$$(\bar{x} - \bar{y}) - z_\gamma \sqrt{\frac{\sigma_x^2}{m} + \frac{\sigma_y^2}{n}} \le \Delta \le (\bar{x} - \bar{y}) + z_\gamma \sqrt{\frac{\sigma_x^2}{m} + \frac{\sigma_y^2}{n}}$$

or, if $\sigma_x = \sigma_y = \sigma$, by

19.3b
$$(\bar{x} - \bar{y}) - z_\gamma \sigma \sqrt{\frac{1}{m} + \frac{1}{n}} \le \Delta \le (\bar{x} - \bar{y}) + z_\gamma \sigma \sqrt{\frac{1}{m} + \frac{1}{n}}.$$

The hypothesis $\Delta = \Delta_0$ is tested by means of the z-statistic

19.4a
$$z = \frac{(\bar{x} - \bar{y}) - \Delta_0}{\sqrt{\sigma_x^2/m + \sigma_y^2/n}}$$

or, if $\sigma_x = \sigma_y = \sigma$, by

19.4b
$$z = \frac{(\bar{x} - \bar{y}) - \Delta_0}{\sigma\sqrt{1/m + 1/n}}.$$

EXAMPLE 19.5a

Let us consider again the two sets of examination scores in (14.1):

x-sample: 72, 79, 93, 91, 70, 95, 82, 80, 74, 86
y-sample: 78, 66, 65, 84, 69, 73, 71, 75, 68, 90, 76

Let us assume also that from past experience it is known that scores of this type are approximately normally distributed with standard deviation 10. As in Chapter 14 we want to test the hypothesis $\mu_x = \mu_y$ (or equivalently, $\Delta = 0$) and also find a 99% confidence interval for Δ. The test statistic (19.4b) becomes

$$z = \frac{82.2 - 74.1}{10\sqrt{\frac{1}{10} + \frac{1}{11}}} = 1.85,$$

indicating acceptance of the hypothesis $\Delta = 0$ at the 5% significance level. For the confidence interval in (19.3b) we need $z_\gamma \sigma \sqrt{1/m + 1/n} = 2.576 \times 10\sqrt{\frac{1}{10} + \frac{1}{11}} = 11.3$, so that the confidence interval becomes

$$(82.2 - 74.1) - 11.3 \le \Delta \le (82.2 - 74.1) + 11.3$$

or

$$-3.2 \le \Delta \le 19.4.$$

144 **Unknown but Equal Population Standard Deviations**
The case when σ_x and σ_y are unknown is somewhat more complicated than the corresponding situation in Chapter 18. Let us assume first that σ_x and σ_y both have the same unknown value σ. An estimate of σ based on both sets of observations is given by

19.6
$$s = \sqrt{\frac{\sum_i (x_i - \bar{x})^2 + \sum_j (y_j - \bar{y})^2}{m + n - 2}}.$$

If we use this estimate s in place of σ in the z-statistic (19.4b) we obtain the statistic

19.7
$$t = \frac{(\bar{x} - \bar{y}) - \Delta_0}{s\sqrt{1/m + 1/n}},$$

which has a t-distribution with $m + n - 2$ degrees of freedom. Correspondingly, the confidence interval in (19.3b) is converted to

19.8 $\quad (\bar{x} - \bar{y}) - t_\gamma s \sqrt{1/m + 1/n} \le \Delta \le (\bar{x} - \bar{y}) + t_\gamma s \sqrt{1/m + 1/n}.$

EXAMPLE
19.5b

We use the same data as in Example 19.5a. Past experience may have told us that it is appropriate to consider test scores to be normally distributed with the same standard deviation in both cases. However we may feel that we do not have sufficient information to assume a definite value for the common standard deviation. We can then use statistic (19.7) to test the hypothesis $\Delta = 0$. We compute $\sum_i (x_i - \bar{x})^2 = 707.6$ and $\sum_j (y_j - \bar{y})^2 = 592.9$, so that $s = \sqrt{(707.6 + 592.9)/19} = 8.27$. Then

$$t = \frac{82.2 - 74.1}{8.27\sqrt{\frac{1}{10} + \frac{1}{11}}} = 2.24.$$

With 19 degrees of freedom the result is significant at the .05-level but not at the .01-level. For a 99% confidence interval we find $t_\gamma s \sqrt{1/m + 1/n} = 2.86 \times 8.27 \sqrt{\frac{1}{10} + \frac{1}{11}} = 10.3$ giving the confidence interval

$$8.1 - 10.3 \le \Delta \le 8.1 + 10.3$$

or

$$-2.2 \le \Delta \le 18.4.$$

The second interval based on t happens to be shorter than the first interval based on z. This is due to the fact that the sample estimate s of σ used in the computations for the second interval turned out to be somewhat smaller than the population value $\sigma = 10$ that we were willing to use when finding the first interval. Since for a given confidence coefficient γ we have $z_\gamma < t_\gamma$, on the average the interval based on z is shorter than the interval based on t (provided we use the correct value for σ).

145 *Unknown and Possibly Unequal Population Standard Deviations* We now turn to the case when σ_x and σ_y are unknown and may not be assumed to be equal. The corresponding z-statistic (19.4a) suggests the use of the test statistic

$$t^* = \frac{(\bar{x} - \bar{y}) - \Delta_0}{\sqrt{s_x^2/m + s_y^2/n}}.$$

If m and n are both large, t^* is approximately normally distributed and the hypothesis $\Delta = \Delta_0$ can be tested by referring t^* to Table B of the normal distribution. However the normal approximation is unsatisfactory if either or both of m and n are small. For this case, the following procedure has been suggested. Let $t_{x\gamma}$ be the appropriate critical value of the t-distribution with $m - 1$ degrees of freedom; $t_{y\gamma}$, that of the t-distribution with $n - 1$ degrees of freedom. Let $w_x = s_x^2/m$; $w_y = s_y^2/n$. The critical value for the statistic t^* is then

$$t_\gamma^* = \frac{w_x t_{x\gamma} + w_y t_{y\gamma}}{w_x + w_y}.$$

If $m = n$, $t_\gamma^* = t_{x\gamma} = t_{y\gamma}$, the critical value of the t-distribution with $m - 1 = n - 1$ degrees of freedom.

A confidence interval for Δ is given by

$$(\bar{x} - \bar{y}) - t_\gamma^* \sqrt{s_x^2/m + s_y^2/n} \leq \Delta \leq (\bar{x} - \bar{y}) + t_\gamma^* \sqrt{s_x^2/m + s_y^2/n}.$$

EXAMPLE 19.9 In a sample of size 10 we have found $\bar{x} = 80.6$ and $s_x^2 = 96.8$. In a second sample of size 15 we have found $\bar{y} = 62.3$ and $s_y^2 = 35.6$. We want to test the hypothesis $\Delta = \mu_x - \mu_y = 10$. We easily find

$$t^* = \frac{(80.6 - 62.3) - 10}{\sqrt{96.8/10 + 35.6/15}} = 2.39,$$

which for a signifiance level of .05 is greater than either $t_{x\gamma} = 2.26$ or $t_{y\gamma} = 2.14$. We can reject the hypothesis $\Delta = 10$ at significance level .05 even without computing the exact value of t_γ^*, since t_γ^* must lie between $t_{x\gamma}$ and $t_{y\gamma}$.

We do need t_γ^* to find a confidence interval for Δ. Since

$$w_x = 96.8/10 = 9.68 \quad \text{and} \quad w_y = 35.6/15 = 2.37,$$

$t_\gamma^* = (9.68 \times 2.26 + 2.37 \times 2.14)/(9.68 + 2.37) = 2.24$. Then $t_\gamma^* \sqrt{s_x^2/m + s_y^2/n} = 2.24\sqrt{9.68 + 2.37} = 7.8$, and a 95% confidence interval for Δ is given by

$$(80.6 - 62.3) - 7.8 \leq \Delta \leq (80.6 - 62.3) + 7.8$$

or

$$10.5 \leq \Delta \leq 26.1.$$

146 The complications arising from potentially different standard deviations when comparing the means of two populations can be avoided by designing an experiment that furnishes observations in pairs, each pair containing one observation from the first population and one observation from the second population. As we have seen, such an experiment is analyzed by considering differences $d_j = x_j - y_j$, involving only the standard deviation of the d's and not the standard deviations of the x's and y's separately.

A Test for Equality of Variances

147   It is sometimes desirable to test the hypothesis that two normal populations have the same standard deviation. Using our customary notation, we want to find out whether $\sigma_x = \sigma_y$. Until now, mostly for intuitive reasons, we have used the standard deviation as a measure of variability. However we have pointed out before that in order to find the standard deviation of a sample (or of a population) it is necessary to compute the variance first. Thus it would have been equally possible to formulate earlier statements involving variability of a sample or of a population in terms of variances. It is more convenient, and certainly more customary, to formulate the problems to be discussed here and in the next chapter in terms of variances.

We therefore want to find out how we can test the hypothesis H_0 that two normal populations have the same variance,

$$H_0: \sigma_x^2 = \sigma_y^2.$$

As in Sections 143–145, we assume that we have two independent samples $x_1, \ldots, x_m$ and $y_1, \ldots, y_n$. We know that s_x^2 is an estimate of σ_x^2 and that s_y^2 is an estimate of σ_y^2. As a consequence we would be inclined to reject H_0 only if s_x^2 and s_y^2 are sufficiently different. Of course, this intuitive statement has to be made more precise. This is done as follows. We compute the ratio of sample variances

19.10
$$F = s_x^2/s_y^2.$$

As long as F is in the neighborhood of 1, we accept H_0. If F deviates significantly from 1, we reject H_0. As with other test procedures, a significant deviation is one that occurs due to chance with probability at most α when H_0 is true. In particular, if we are interested in the one-sided alternative $\sigma_x^2 > \sigma_y^2$, we want to reject H_0 only if the value of F is sufficiently large, say at least F_α. The appropriate value of F_α is read from Table I. This table depends on three quantities: the significance level to be achieved, the number of degrees of freedom of the sample variance in the numerator, and the number of degrees of freedom of the sample variance in the denominator. For our present problem, the latter two quantities are $m - 1$ and $n - 1$, respectively. In Table I the number of degrees of freedom for the numerator is listed in the top row; the number of degrees of freedom for the denominator, in the column on the left.

Next suppose that we are interested in the one-sided alternative $\sigma_x^2 < \sigma^2$. In this case we would want to reject H_0 if $F = s_x^2/s_y^2$ is sufficiently small. While it is of course possible to compute tables of lower critical F-values, there is really no need for such a table. Indeed we can simply rewrite our alternative as $\sigma_y^2 > \sigma_x^2$ and use as our test statistic

$$F' = s_y^2/s_x^2,$$

which now has $n - 1$ numerator degrees of freedom and $m - 1$ denominator degrees of freedom. We reject H_0 if $F' > F_\alpha$, where F_α is found as before (with reversed degrees of freedom).

148   This still leaves the two-sided alternative $\sigma_x^2 \neq \sigma_y^2$. As in previous tests against two-sided alternatives, the appropriate critical region consists of two parts, large and small values of F (or F', if we prefer). Since Table I does not provide us with lower critical values, we combine the F- and F'-tests as follows. We reject the hypothesis $\sigma_x^2 = \sigma_y^2$ against the alternative $\sigma_x^2 \neq \sigma_y^2$ if the larger of

$$F = s_x^2/s_y^2 \text{ and } F' = s_y^2/s_x^2$$

is greater than the appropriate critical value read from Table I. In practice we compute only F or F', whichever has the larger sample variance in the numerator. This test has significance level $\alpha'' = 2\alpha$, where α is the upper tail probability associated with the critical value being used.

EXAMPLE   We want to test the hypothesis that the x- and y-scores considered
19.11   in Example 19.5a have equal variance against the two-sided alternative of unequal variances. Since for the two samples s_x^2 is greater than s_y^2, the appropriate test statistic is $F = s_x^2/s_y^2 = 78.6/59.3 = 1.33$. With 9 numerator degrees of freedom and 10 denominator degrees of freedom, the .05 upper critical value is 3.02. This value corresponds to a two-sided significance level of .10. Clearly our result is not significant.

PROBLEMS   1 Show that $\bar{d} = \bar{x} - \bar{y}$.

2 Assume that pulse rates are normally distributed, and consider the data of Problem 2, Chapter 13.
   a. Find a confidence interval for the mean pulse rate before exercise.
   b. Find a confidence interval for the mean increase in pulse rate due to exercise.
   c. Is the hypothesis that the mean increase is 20 consistent with the data?

3 Solve Problem 4, Chapter 13 by the methods of this chapter. What additional assumptions, if any, do you have to make?

4 In a sample of size 16 from a normal population you have found $\sum_i x_i = 1205$ and $\sum_i x_i^2 = 92411$. In an independent sample of size 22 from another normal population you have found $\sum_j y_j = 1767$ and $\sum_j y_j^2 = 143921$.

a. Test the hypothesis $\mu_x = \mu_y$.
b. Find a confidence interval for $\Delta = \mu_x - \mu_y$ having confidence coefficient .98.
c. Test the hypothesis $\sigma_x^2 = \sigma_y^2$.

5 Solve Problem 6, Chapter 14 by the methods of this chapter. What additional assumptions, if any, do you have to make?

6 The GRE scores of 10 students in Department X are: 646, 655, 582, 504, 407, 779, 696, 830, 772, 745. The GRE scores of 12 students in Department Y are: 649, 837, 690, 669, 634, 698, 548, 615, 685, 659, 657, 805. Would you say that the two departments differ in their admission policy. Support your statements by appropriate statistical arguments.

7 For each of the following problems decide whether it may be solved by the methods of this chapter. If your answer is no, state why not. If your answer is yes, state the appropriate procedure and required assumptions, if any.
    a. Problem 3, Chapter 13;
    b. Problem 4, Chapter 13;
    c. Problem 1, Chapter 14;
    d. Problem 4, Chapter 14;
    e. Problem 5, Chapter 14;
    f. Problem 8, Chapter 14.

8 Using the information of Example 19.9, test the hypothesis $\sigma_x^2 = \sigma_y^2$ against the alternative $\sigma_x^2 > \sigma_y^2$.

20

---

k-Sample Problems for Normal Populations

The One-Way Classification

149 In its simplest form the problem to be considered in this chapter can be stated as follows. We have k independent samples, one each from k normal populations with means $\mu_1, \ldots, \mu_k$, and want to test the hypothesis

20.1
$$H_0: \mu_1 = \cdots = \mu_k$$

or find confidence intervals for differences of means

$$\mu_g - \mu_h, \ 1 \le g \ne h \le k,$$

as well as more complicated expressions in the μ's.

Throughout this chapter we assume that the k normal populations have the same (unknown) variance σ^2. Violation of this assumption in practice may seriously invalidate the results.

Essentially our approach to finding a test statistic for the hypothesis (20.1) is to generalize the idea underlying the test statistic (19.7) for the corresponding two-sample problem. In (19.7) the difference $\bar{x} - \bar{y}$ is a measure of how far apart the population means μ_x and μ_y are. However since $\bar{x}$ and $\bar{y}$ are subject to sampling fluctuations, we need a standard with which to compare the difference $\bar{x} - \bar{y}$. The denominator in (19.7) serves as such a standard. Similarly, in the present case we find a quantity (to be called MS_A) that is a measure of how far apart the population means $\mu_1, \ldots, \mu_k$ are. Again because of sampling fluctuations, this measure has to be compared with an appropriate standard (to be called MS_W).

150 *Analysis of Variance* A precise and somewhat elaborate notation is helpful when discussing our problem. The jth observation in the ith sample is denoted by x_{ji}, $i = 1, \ldots, k; j = 1, \ldots, n_i$. We have the sum of the observations in the ith sample

$$T_i = \sum_j x_{ji};$$

the mean of the observations in the ith sample

$$\bar{x}_i = T_i/n_i;$$

the sum of all $N = n_1 + \cdots + n_k$ observations

$$T = \sum_i T_i = \sum_i \sum_j x_{ji};$$

and the overall sample mean

$$\bar{x} = T/N.$$

The analysis of variance, which is basic for the methods of this chapter, as well as the next, receives its name from the fact that the so-called total sum of squares

$$SS_T = \sum_i \sum_j (x_{ji} - \bar{x})^2,$$

which characterizes the spread of all N observations about the overall mean $\bar{x}$, is written as the sum of two or more constituent sums of squares, each of which characterizes a different factor that contributes to the overall variability of the observations. In the simplest case we have two components,

20.2
$$SS_T = SS_W + SS_A$$

where

$$SS_W = \sum_i \sum_j (x_{ji} - \bar{x}_i)^2$$

and

$$SS_A = \sum_i \sum_j (\bar{x}_i - \bar{x})^2 = \sum_i n_i (\bar{x}_i - \bar{x})^2.$$

Since for every i, $\sum_j (x_{ji} - x_i)^2$ measures variability within the ith sample, SS_W is called the *within* sum of squares, while SS_A is usually called the *among* or *treatment* sum of squares.

151 In order to derive (20.2) we write

$$x_{ji} - \bar{x} = (x_{ji} - \bar{x}_i) + (\bar{x}_i - \bar{x}),$$

that is, we express the deviation of x_{ji} from the overall mean $\bar{x}$ as the sum of two deviations: (i) the deviation of x_{ji} from the mean x_i of all observations in the same sample, and (ii) the deviation of the particular sample mean $\bar{x}_i$ from the overall mean $\bar{x}$. Substitution gives

$$SS_T = \sum_i \sum_j [(x_{ji} - \bar{x}_i) + (\bar{x}_i - \bar{x})]^2$$
$$= \sum_i \sum_j (x_{ji} - \bar{x}_i)^2 + \sum_i \sum_j (\bar{x}_i - \bar{x})^2,$$

since the sum of the crossproduct terms vanishes. This is most easily seen as follows

$$\sum_i\sum_j(x_{ji} - \bar{x}_i)(\bar{x}_i - \bar{x}) = \sum_i(\bar{x}_i - \bar{x})[\sum_j(x_{ji} - \bar{x}_i)]$$

and

$$\sum_j(x_{ji} - \bar{x}_i) = T_i - n_i\bar{x}_i = 0$$

for every *i*.

152 Associated with each sum of squares is an appropriate number of degrees of freedom: $N - 1$ for SS_T, $\sum_i(n_i - 1) = N - k$ for SS_W, $(N - 1) - (N - k) = (k - 1)$ for SS_A. By dividing SS_W and SS_A by the respective number of degrees of freedom, we obtain the *mean squares*

$$MS_W = SS_W/(N - k) \text{ and } MS_A = SS_A/(k - 1).$$

These mean squares have the following significance for our problem. MS_W is an estimate of the unknown variance σ^2 common to all *k* normal populations, whether H_0 is true or not. On the other hand, the treatment mean square MS_A is an estimate of

$$\sigma_A^2 = \sigma^2 + \frac{1}{k - 1}\sum_i n_i(\mu_i - \mu)^2,$$

where $\mu = \sum_i n_i\mu_i/N$. Thus MS_W measures fluctuation caused by sampling, while MS_A measures in addition how far apart $\mu_1, \ldots, \mu_k$ are.

When H_0 is true, both MS_A and MS_W are estimates of the same quantity σ^2, while under the alternative hypothesis, MS_A is an estimate of a quantity that is greater than σ^2. We can then reformulate our problem as a test of the hypothesis

$$\sigma_A^2 = \sigma^2,$$

involving two unknown variances σ^2 and σ_A^2, against the alternative

$$\sigma_A^2 > \sigma^2.$$

We saw in Chapter 19 that this hypothesis is tested by means of the *F*-statistic

20.3
$$F = MS_A/MS_W$$

with $k - 1$ degrees of freedom for the numerator and $N - k$ degrees of freedom for the denominator, using a one-sided test with large values of *F* indicating rejection of H_0.

The *F*-test (20.3) is the normal theory equivalent of the *H*-test of Chapter 15. Indeed, the methods of Chapter 15 are sometimes referred to as nonparametric analysis of variance procedures.

153 *Computations* The expressions that we have used to define the sums of squares SS_T, SS_W, and SS_A are not very suitable for computations with actual data. Instead we compute

20.4
$$C = T^2/N, \quad D = \sum_i\sum_j x_{ji}^2, \quad \text{and} \quad E = \sum_i T_i^2/n_i$$

to obtain

20.5a	$SS_T = D - C,$
20.5b	$SS_W = D - E,$
20.5c	$SS_A = E - C.$

Indeed, the derivation of (20.5a) is analogous to that in (18.4). With the help of the same expression, we find that

$$\sum_j (x_{ji} - \bar{x}_i)^2 = \sum_j x_{ji}^2 - T_i^2/n_i,$$

and therefore

$$SS_W = \sum_i \left[\sum_j (x_{ji} - \bar{x}_i)^2\right] = \sum_i \sum_j x_{ji}^2 - \sum_i T_i^2/n_i = D - E.$$

Finally, from (20.2)

$$SS_A = SS_T - SS_W = (D - C) - (D - E) = E - C.$$

It is customary to exhibit all available information in the Analysis of Variance Table 20.6:

TABLE 20.6

Analysis of Variance Table: One-Way Classification

source of variation	sum of squares SS	degrees of freedom df	mean square MS = SS/df
among samples	$SS_A = E - C$	$k - 1$	$MS_A = SS_A/(k - 1)$
within samples	$SS_W = D - E$	$N - k$	$MS_W = SS_W/(N - k)$
total	$SS_T = D - C$	$N - 1$	

154 As an illustration of the required computations we look once more at the data in Table 15.3a representing stopping distances for five brands of automobile tires. For the sake of convenience the data are repeated in Table 20.7. The present analysis assumes that the

TABLE 20.7

	brand of tire			
A	B	C	D	E
151	157	135	147	146
143	158	146	174	171
159	150	142	179	167
152	142	129	163	145
156	140	139	148	147
			165	166

data represent samples from normal populations having identical variance. We want to test the hypothesis

$$\mu_A = \mu_B = \mu_C = \mu_D = \mu_E.$$

Computations proceed as follows:

	A	B	C	D	E	
T_i	761	747	691	976	942	$T = 4117$
n_i	5	5	5	6	6	$N = 27$

$$C = T^2/N = 4117^2/27 = 627766.3,$$
$$D = \sum_i \sum_j x_{ji}^2 = 151^2 + \cdots + 166^2 = 631775,$$
$$E = \sum_i T_i^2/n_i = 761^2/5 + \cdots + 942^2/6 = 629578.9,$$
$$SS_T = D - C = 4008.7,$$
$$SS_A = E - C = 1812.6,$$
$$SS_W = D - E = SS_T - SS_A = 2196.1.$$

The Analysis of Variance Table 20.6 now becomes:

source of variation	SS	df	MS
among samples	1812.6	4	453.2
within samples	2196.1	22	99.8
total	4008.7	26	

Finally, $F = 453.2/99.8 = 4.54$. This value of F surpasses the critical value $F_{.01} = 4.31$ for 4 numerator and 22 denominator degrees of freedom. Thus we reject the null hypothesis and conclude that the average braking distance for the 5 brands of tires is presumably not the same.

155 We have developed the F-test as a test of the hypothesis that k normal populations have identical means, where the number of populations is greater than two. Actually our procedure can also be used for the comparison of the means of two normal populations. It is not difficult to show that the square of the t-statistic (19.7) (with $\Delta_0 = 0$) for two independent samples equals the F-statistic of this chapter. Since large values of F, which suggest rejection of the hypothesis of equal means, result from either large positive or large negative values of t, for $k = 2$ the F-test should be used only as a test against the two-sided alternative $\mu_x \neq \mu_y$.

Simultaneous Confidence Intervals

156 When we reject the hypothesis that the means $\mu_1, \ldots, \mu_k$ are all equal, it is only natural to ask which means differ from which and by how much. The simplest type of comparison is between pairs of means. Thus we may want to estimate differences like

20.8

$$\mu_1 - \mu_2.$$

On the other hand practical considerations often suggest more complicated comparisons involving more than two means. If the investigator feels that populations 1 and 2 are rather similar, but that population 3 is different, he may be interested in comparing μ_3 with the average of μ_1 and μ_2, that is, find an estimate of

20.9
$$\mu_3 - \tfrac{1}{2}(\mu_1 + \mu_2).$$

The expressions in (20.8) and (20.9) are examples of *contrasts*. More generally, a contrast ψ based on k means $\mu_1, \ldots, \mu_k$ is a linear function of the μ's,

$$\psi = c_1\mu_1 + \cdots + c_k\mu_k,$$

where the constants c_i are completely arbitrary except that $c_1 + \cdots = c_k = 0$. For example, in (20.8) we have $c_1 = 1$, $c_2 = -1$, $c_3 = \cdots = c_k = 0$. For the contrast (20.9), $c_1 = c_2 = -\tfrac{1}{2}$, $c_3 = 1$, $c_4 = \cdots = c_k = 0$. Contrasts of the type (20.8) are often called *simple* contrasts.

157
Since the sample means $\bar{x}_i$ are estimates of the population means μ_i, a natural point estimate of the contrast ψ is

$$\hat{\psi} = c_1\bar{x}_1 + \cdots + c_k\bar{x}_k.$$

While point estimates have their legitimate uses, estimation by confidence interval is usually more informative. We therefore want to find out how to construct confidence intervals for contrasts. It is possible to generalize the confidence interval approach of Chapter 19 using an appropriate t-statistic. But the present problem is complicated by two additional factors. We are rarely interested in just one contrast at a time. Thus we may want to have confidence intervals for both $\mu_1 - \mu_2$ and $\mu_3 - \tfrac{1}{2}(\mu_1 + \mu_2)$ as well as other contrasts. The approach of Chapter 19 allows us to assign confidence coefficients separately to each interval that we compute. But it leaves unanswered the question about the degree of overall confidence we can have in the correctness of all confidence statements we make. In addition, certain contrasts may come to our attention only when we look at the results of our experiment and note certain extreme differences among sample means. The following remarkable theorem resolves both difficulties for us:

THEOREM 20.10
Scheffé's Theorem *Let $S^2 = (k - 1)F_\alpha$, where F_α is the significance point for the F-test of the hypothesis $\mu_1 = \cdots = \mu_k$. For any particular contrast $\psi = c_1\mu_1 + \cdots + c_k\mu_k$, let $\hat{\sigma}_{\hat{\psi}}^2 = MS_W \times \sum_i c_i^2/n_i.$* Then the overall confidence we can have in*

*We have $V(\hat{\psi}) = (c_1^2/n_1 + \cdots + c_k^2/n_k)\sigma^2$, so that $\hat{\sigma}_{\hat{\psi}}^2$ is an estimate of $V(\hat{\psi})$ using the within mean square MS_W to estimate the unknown population variance σ^2.

all possible statements of the form

20.11
$$\hat{\psi} - S\hat{\sigma}_{\hat{\psi}} \le \psi \le \hat{\psi} + S\hat{\sigma}_{\hat{\psi}}$$

is $\gamma = 1 - \alpha$.

The fact that the overall confidence coefficient γ refers to any and all statements of type (20.11) allows us to look at our data and actually select contrasts suggested by such a look. Furthermore, since in any particular case we are interested in only a limited number of contrasts, the true overall confidence coefficient is actually at least γ.

158 We illustrate simultaneous confidence intervals using the tire data in Table 20.7. We are interested in confidence intervals for all possible differences of pairs of means. But suppose we notice that brands A and C are radial tires and brands B, D, and E are conventional tires. We then may also want to compare the braking performance of radial tires with that of conventional tires.

We start with the computation of sample means. It is convenient to arrange the sample means in increasing (or decreasing) order:

brand	C	B	A	E	D
sample mean	138.2	149.4	152.2	157.0	162.7
sample size	5	5	5	6	6

We need $S\hat{\sigma}_{\hat{\psi}} = \sqrt{(k-1)F_\alpha \times MS_W \sum_i c_i^2/n_i}$. The quantity $(k-1)F_\alpha \times MS_W$ remains the same for every contrast. The quantity $\sum_i c_i^2/n_i$ depends on the specific type of contrast in which we are interested. Using the computations in Section 154 and choosing $\alpha = .25$, we find $(k-1)F_{.25} \times MS_W = 4 \times 1.454 \times 99.82 = 580.55$. For contrasts of the type $\mu_g - \mu_h$ we have $\sum_i c_i^2/n_i = 1/n_g + 1/n_h$, leading to the following numerical results:

n_g	n_h	$1/n_g + 1/n_h$	$S\hat{\sigma}_{\hat{\psi}}$
5	5	.400	15.2
5	6	.367	14.6
6	6	.333	13.9

Some examples of confidence intervals for contrasts of type $\mu_g - \mu_h$ are:

$$(138.2 - 157.0) - 14.6 \le \mu_C - \mu_E \le (138.2 - 157.0) + 14.6$$

or

$$-33.4 \le \mu_C - \mu_E \le -4.2.$$

Similarly,

$$(152.2 - 138.2) - 15.2 \le \mu_A - \mu_C \le (152.2 - 138.2) + 15.2$$

or

$$-1.2 \leq \mu_A - \mu_C \leq 29.2.$$

Also

$$-27.9 \leq \mu_B - \mu_D \leq 1.3.$$

Radial and conventional tires are compared by means of the contrast

$$\psi = \tfrac{1}{2}(\mu_A + \mu_C) - \tfrac{1}{3}(\mu_B + \mu_D + \mu_E),$$

for which

$$\sum_i \frac{c_i^2}{n_i} = \frac{1}{4}\left(\frac{1}{5} + \frac{1}{5}\right) + \frac{1}{9}\left(\frac{1}{5} + \frac{1}{6} + \frac{1}{6}\right) = .159,$$

so that $S\hat{\sigma}_{\hat{\psi}} = \sqrt{580.55 \times .159} = 9.6$. Since

$$\hat{\psi} = \tfrac{1}{2}(152.2 + 138.2) - \tfrac{1}{3}(149.4 + 157.0 + 162.7) = -11.2,$$

we find the confidence interval

$$-11.2 - 9.6 \leq \psi \leq -11.2 + 9.6$$

or

$$-20.8 \leq \psi \leq -1.6.$$

159 *Confidence Intervals and Tests of Hypotheses*   We have remarked on several occasions that a confidence interval consists of all acceptable parameter values. If ψ is a contrast, we can test the hypothesis $\psi = 0$ by finding out whether or not the confidence interval (20.11) for ψ contains the origin. Since this confidence interval does not contain the origin if either $\hat{\psi} - S\hat{\sigma}_{\hat{\psi}} > 0$ or if $\hat{\psi} + S\hat{\sigma}_{\hat{\psi}} < 0$, the hypothesis $\psi = 0$ should be rejected if

20.12
$$|\hat{\psi}| > S\hat{\sigma}_{\hat{\psi}}.$$

Looking at all possible contrasts we can establish an interesting relationship between confidence intervals for contrasts on the one hand and the analysis of variance test (20.3) of the hypothesis $H_0: \mu_1 = \cdots = \mu_k$ on the other hand. If ψ is any contrast and if H_0 is true,

$$\psi = c_1\mu_1 + \cdots + c_k\mu_k = (c_1 + \cdots + c_k)\mu_1 = 0,$$

since according to the definition of a contrast we have $c_1 + \cdots + c_k = 0$. If then H_0 is true, we can make the following two statements. First, however many contrasts we decide to consider, with probability $\gamma = 1 - \alpha$ *all* confidence intervals (20.11) contain the origin. Secondly, if $F = MS_A/MS_W$ is the analysis of variance test statistic, with the same probability $1 - \alpha$ we have $F \leq F_\alpha$, indicating acceptance of H_0. Indeed it can be shown that these two statements are equivalent in the following sense. If $F \leq F_\alpha$, all confidence intervals (20.11) contain the origin. On the other hand if $F > F_\alpha$, then there exist some confidence intervals (20.11) that do not contain the origin.

Probability Models

160 ***One-Way Classification*** For a better understanding of the remainder of this chapter, as well as the next, it is helpful to express our basic assumptions in precise mathematical form. So far we have simply stated that x_{ji} represents the jth observation from a normal population with mean μ_i and variance σ^2. This statement can be formalized by writing

$$x_{ji} = \mu_i + e_{ji}, \; N(e_{ji}; 0, \sigma^2)$$

indicating that the observed quantity x_{ji} deviates from the population mean μ_i by the amount e_{ji}, where e_{ji} is assumed to be normal with mean 0 and variance σ^2. Indeed we may go one step further by writing

20.13
$$\mu_i = \mu + \alpha_i,$$

where by definition $\mu = \sum_i n_i \mu_i / N$ is a weighted average of the μ_i. We then have the probability model

20.14
$$x_{ji} = \mu + \alpha_i + e_{ji}, \; N(e_{ji}; 0, \sigma^2)$$

so that x_{ji} appears as the sum of three quantities: an overall mean μ; the deviation α_i of μ_i from μ; and the normal deviation e_{ji} of the observation x_{ji} from its population mean μ_i. If the null hypothesis is true, we have

$$\mu_1 = \cdots = \mu_k = \mu,$$

and therefore

$$\alpha_i = \mu_i - \mu = 0, \; i = 1, \ldots, k.$$

The analysis of variance test (20.3) thus becomes a test of the hypothesis

$$\alpha_1 = \cdots = \alpha_k = 0$$

against the alternative that not all α_i vanish, regardless of the true values of the parameters μ and σ^2 appearing in model (20.14). We note in passing that in view of (20.13),

$$\sigma_A^2 = \sigma^2 + \frac{1}{k-1} \sum_i n_i \alpha_i^2.$$

Under H_0 this reduces to $\sigma_A^2 = \sigma^2$.

161 ***Randomized Blocks*** We have mentioned that the analysis of variance test of Section 152 is the normal theory equivalent of the Kruskal-Wallis test. We may then ask, what is the normal theory equivalent of the Friedman test for randomized blocks?

Let us first of all introduce a suitable mathematical model. Now we write x_{ji} for the observation in the jth block that corresponds to treatment i. As before, i varies from 1 to k; but j varies from 1 to n corresponding to the n blocks. Our first encounter with "blocked" observations occurred in Chapter 13 in connection with the comparison of two suntan preparations. At that time we remarked that by applying both suntan lotions to the back of the same test subject, we hoped to reduce variability among measurements caused by differences in the skin sensitivity of different test subjects. More formally let x_{j1} and x_{j2} represent the sunburn measurements for the jth test subject corresponding to lotions 1 and 2. (We use a somewhat different notation from that in Chapter 13, but we have in mind the generalization from 2 to k treatments.) The following suggests itself as a possible model:

$$x_{j1} = \mu + \alpha_1 + \beta_j + e_{j1}$$
$$x_{j2} = \mu + \alpha_2 + \beta_j + e_{j2}.$$

Here μ is the overall mean of all conceivable sunburn measurements; α_1 and α_2 indicate deviations from μ caused by the protective abilities of lotions 1 and 2, respectively; the β_j indicate deviations from μ due to differing skin sensitivity of the various test subjects; finally, e_{j1} and e_{j2} reflect all other sources of variability such as differing exposure to the sun, previous exposure, possible variations in exposure time, and so on. The reason for basing our analysis on differences of paired observations now becomes clear. We have

$$d_j = x_{j1} - x_{j2} = (\alpha_1 - \alpha_2) + e_j$$

where $e_j = e_{j1} - e_{j2}$ measures the difference in uncontrolled factors between the side protected by lotion 1 and the side protected by lotion 2. The important point to note is that d_j reflects the difference in protective ability of lotions 1 and 2, if such a difference exists, but does not depend on the block factor β_j.

162 Our model for the suntan lotion experiment suggests the following generalization:

$$x_{ji} = \mu + \alpha_i + \beta_j + e_{ji}, \ N(e_{ji}; 0, \sigma^2)$$
$$i = 1, \ldots, k; \ j = 1, \ldots, n.$$

Here μ is the overall mean of the type of measurements we are performing; $\alpha_1, \ldots, \alpha_k$ characterize the specific effects of the treatments under investigation; β_j is a specific effect for the jth block; finally, the e_{ji} take care of all other uncontrolled factors that influence x_{ji}. In the present analysis, it is assumed that the e_{ji} are independent normal variables with means 0 and (unknown) variance σ^2. The present setup can also be described by saying that the x_{ji} are normal variables with means

20.15
$$\mu_{ji} = \mu + \alpha_i + \beta_j$$

and constant variance σ^2.

163 Model (20.15) allows us to specify more precisely the statistical problems we want to solve. The earlier statement that all k samples have come from one and the same population is now formulated more precisely as

20.16

$$\alpha_1 = \cdots = \alpha_k = 0.$$

We want to derive a test for hypothesis (20.16). Alternatively, if (20.16) is false, we may want to find confidence intervals for various contrasts

$$\psi = c_1\alpha_1 + \cdots + c_k\alpha_k,$$

where the c_i's are given constants with $c_1 + \cdots + c_k = 0$.

From a purely formal point of view, the β_j's appear in model (20.15) symmetrically to the α_i's. Thus any test of the hypothesis (20.16) suggests a corresponding test of the hypothesis

20.17

$$\beta_1 = \cdots = \beta_n = 0.$$

In practice, if the experimenter uses a randomized block design because he suspects the presence of block to block differences, there is not much point in testing (20.17). We shall see, however, that under a different experimental setup, the hypothesis (20.17) may require testing.

Analysis of Variance for Randomized Blocks

164   Let x_{ji} be the observation on treatment i occurring in block j. We define the following sums and averages:

$$T_i = \Sigma_j x_{ji} = \text{sum of all observations for } i\text{th treatment}$$

$$\bar{x}_{.i} = \frac{1}{n}\, T_i = \text{mean of all observations for } i\text{th treatment}$$

$$B_j = \Sigma_i x_{ji} = \text{sum of all observations in } j\text{th block}$$

$$\bar{x}_{j.} = \frac{1}{k}\, B_j = \text{mean of all observations in } j\text{th block}$$

$$T = \Sigma_i T_i = \Sigma_j B_j = \text{sum of all observations}$$

$$\bar{x} = \frac{1}{nk}\, T = \text{mean of all observations.}$$

(The dot notation used above in $\bar{x}_{.i}$ and $\bar{x}_{j.}$ is very useful. A dot in place of a subscript indicates that we have averaged those observations whose subscript has been replaced by the dot. According to this convention we should write $\bar{x}_{..}$ in place of $\bar{x}$, but we shall use the simpler $\bar{x}$, since no confusion can arise.)

We now write

$$x_{ji} - \bar{x} = (\bar{x}_{.i} - \bar{x}) + (\bar{x}_{j.} - \bar{x}) + (x_{ji} - \bar{x}_{.i} - \bar{x}_{j.} + \bar{x}),$$

and proceeding as in Section 151, obtain the following partition of the total sum of squares:

$$SS_T = \sum_i \sum_j (x_{ji} - \bar{x})^2$$
$$= \sum_i \sum_j (\bar{x}_i - \bar{x})^2 + \sum_i \sum_j (\bar{x}_{j.} - \bar{x})^2$$
$$+ \sum_i \sum_j (x_{ji} - \bar{x}_{.i} - \bar{x}_{j.} + \bar{x})^2$$
$$= SS_A + SS_B + SS_R.$$

SS_A is called the treatment sum of squares as before; SS_B, the block sum of squares; and SS_R, the residual (or error) sum of squares, that is, the part of the total sum of squares that remains after the sum of squares due to treatments and due to blocks has been eliminated. SS_A again has $k - 1$ degrees of freedom and by symmetry, SS_B has $n - 1$ degrees of freedom, leaving $(nk - 1) - (k - 1) - (n - 1) = (n - 1)(k - 1)$ degrees of freedom for SS_R. As before, our information can be consolidated in an analysis of variance table, in which computing formulas for SS_T, SS_A, SS_B, and SS_R are derived as in Section 153:

Analysis of Variance Table: Randomized Blocks

source of variation	sum of squares SS	degrees of freedom df	mean square $MS = SS/df$
treatments	$SS_A = E - C$	$k - 1$	$MS_A = SS_A/(k - 1)$
blocks	$SS_B = F - C$	$n - 1$	$MS_B = SS_B/(n - 1)$
residuals	$SS_R = SS_T - SS_A - SS_B$	$(k - 1)(n - 1)$	$MS_R = SS_R/(k - 1)(n - 1)$
total	$SS_T = D - C$		

As before, we set $C = T^2/nk$, $D = \sum_i \sum_j x_{ij}^2$, and $E = \sum_i T_i^2/n$, and we let $F = \sum_j B_j^2/k$.

165   It can be shown that, regardless of the α_i's and β_j's, MS_R is an estimate of σ^2. On the other hand, MS_A estimates the quantity

20.18
$$\sigma_A^2 = \sigma^2 + \frac{n}{k - 1} \sum_i \alpha_i^2$$

and MS_B estimates the quantity

20.19
$$\sigma_B^2 = \sigma^2 + \frac{k}{n - 1} \sum_j \beta_j^2.$$

Expression (20.18) suggests the following test of the hypothesis (20.16). If (20.16) is true, we have $\sigma_A^2 = \sigma^2$, and this hypothesis is tested using the statistic

$$F = MS_A/MS_R$$

with $k - 1$ numerator and $(k - 1)(n - 1)$ denominator degrees of freedom. Similarly, hypothesis (20.17), when appropriate, is tested using the statistic $F = MS_B/MS_R$ with $n - 1$ numerator and $(k - 1)(n - 1)$ denominator degrees of freedom.

EXAMPLE For the gasoline mileage data in Table 15.5 we have $k = 3$ and
20.20 $n = 5$. Further, $T = 315.3$ and therefore,

$$C = 315.3^2/15 = 6627.61$$
$$D = 20.3^2 + \cdots + 21.4^2 = 6703.87$$
$$\sum_i T_i^2 = 96.8^2 + 109.6^2 + 108.9^2 = 33241.61$$
$$\sum_j B_j^2 = 69.9^2 + 69.0^2 + 58.1^2 + 57.7^2 + 60.6^2 = 20024.27$$

so that

$$E = 33241.61/5 = 6648.32$$
$$F = 20024.27/3 = 6674.76.$$

The analysis of variance table then becomes:

source of variation	SS	df	MS
treatments	20.71	2	10.36
blocks	47.15	4	11.79
residual	8.40	8	1.05
total	76.26		

We test the hypothesis that the three makes of cars deliver identical gasoline mileage by means of the statistic $F = MS_A/MS_R = 10.36/1.05 = 9.87$. This value of F surpasses the critical value 8.65 corresponding to $\alpha = .01$, and we reject our hypothesis.

166 *Confidence Intervals for Contrasts* For the present model we consider contrasts

$$\psi = c_1 \alpha_1 + \cdots + c_k \alpha_k$$

where the c_i are arbitrary constants with $c_1 + \cdots + c_k = 0$. Confidence intervals for all possible contrasts are obtained almost exactly as in Section 157. Since α_i measures the deviation of the effect of the ith treatment from the overall mean μ, α_i is estimated as $\bar{x}_{.i} - \bar{x}$, suggesting that we estimate ψ as

$$\hat{\psi} = c_1(\bar{x}_{.1} - \bar{x}) + \cdots + c_k(\bar{x}_{.k} - \bar{x})$$
$$= c_1\bar{x}_{.1} + \cdots + c_k\bar{x}_{.k} - (c_1 + \cdots + c_k)\bar{x}$$
$$= c_1\bar{x}_{.1} + \cdots + c_k\bar{x}_{.k}.$$

When computing $\hat{\sigma}_{\hat{\psi}}^2$, MS_W of Section 157 is replaced by MS_R with a corresponding change in the denominator degrees of freedom from $N - k$ to $(k - 1)(n - 1)$.

Two-Factor Experiments

167 We have pointed out that in a randomized block design there is little justification for testing the hypothesis (20.17) stating that

$\beta_1 = \cdots = \beta_n$. Presumably the experimenter chose the randomized block design because he thought that not all β's were equal. But model (20.15) is also appropriate for experiments other than those with randomized block designs. The subscripts i and j may refer to two different types of treatments, similar to the two-way classifications in Chapter 11. To return to an earlier example, classification 1 may again refer to various ways of teaching a school subject; classification 2 may refer to different types of schools. Here the question of interest is to find out not only whether the different teaching methods produce different results, but also whether there are differences from one type of school to another. The second question can be answered by testing the hypothesis (20.17) and possibly finding confidence intervals for contrasts of the type

$$\psi = d_1\beta_1 + \cdots + d_n\beta_n, \ d_1 + \cdots + d_n = 0.$$

Experiments of the type just discussed are often called *two-factor* experiments. Two- and several-factor experiments represent important topics for statistical theory and analysis. However, they take us beyond the scope of an introductory treatment. We only want to mention here that model (20.15) is often too simple for general two-factor experiments. Model (20.15) is referred to as an additive model, since the individual effects of the two factors enter the model as the sum $\alpha_i + \beta_j$, and not in some more complicated fashion. In a more general treatment of two-factor experiments, model (20.15) has to be modified by the addition of *interaction* terms γ_{ji}

$$\mu_{ji} = \mu + \alpha_i + \beta_j + \gamma_{ji},$$

indicating that the specific combination of level i of factor 1 and level j of factor 2 produces a result that cannot be explained simply as $\alpha_i + \beta_j$. In our previous example, it is quite conceivable that a particular teaching method may produce very good results in one type of school but not in another. We then say that type of school and method of teaching interact. In general, in such a situation no one teaching method produces best results. The results also depend on the type of school where the method is used.

Of course, it is possible that a set of blocks designed to reduce variability actually interacts with the treatments under consideration. In such a case, the results of our analysis may be misleading or even invalid. For example in the suntan lotion experiment, a certain lotion may provide very good protection against sunburn for one kind of skin but not for another. Before an experimenter decides to use a randomized block design, he should be reasonably certain that blocks do not interact with treatments.

PROBLEMS 1 For the tire data find simultaneous confidence intervals for all simple contrasts. Show that these confidence intervals imply the same conclusions as the multiple comparison analysis of the same data in Chapter 15.

2 Repeat the simultaneous confidence interval analysis for the tire data given in the book using $\alpha = .10$.

3 Analyze the data of Problem 1, Chapter 15 by the methods of this chapter. In particular,
   a. set up and test a relevant hypothesis,
   b. find confidence intervals for all simple contrasts,
   c. compare diets A and D with diets B and C,
   d. compare diet D with diets A, B, and C.

4 Analyze the data of Problem 6, Chapter 15 by the methods of this chapter. In particular,
   a. set up and test a relevant hypothesis,
   b. find confidence intervals for all simple contrasts,
   c. compare teaching methods 1 and 2 with method 3.

5 For Example 20.20,
   a. find confidence intervals for all simple contrasts,
   b. compare manufacturer G with manufacturers F and C.

6 Among the problems for Chapter 15, are there any problems other than Problems 1 and 6 that can conceivably be solved by the methods of this chapter? If a problem cannot be solved by the methods of this chapter, indicate why not. If you think that a problem can be solved, indicate how.

21

Regression and Correlation

168   In this chapter we consider the normal theory approach to problems encountered in Chapters 11 and 16. In Chapter 11 we dealt with the question of whether or not two characteristics used to classify sample items were related. In Chapter 16 we were concerned with the degree of relationship, if any, of two rankings. Two such rankings may, but need not, be based on sets of measurements involving two continuous variables.

In this chapter we assume that we have pairs of measurements involving two continuous variables conventionally denoted by x and y. We have already encountered several examples. For instance, x may refer to a student's high school grade point average and y, to his college grade point average. Or x may be a person's height and y, his weight. In Chapter 16 we used the Kendall rank correlation coefficient to measure the strength of relationship between x and y. In Section 187 we define a different correlation coefficient. But now we are primarily interested in prediction: If we know the value of one variable, say, x, what can we infer about the other variable?

Two Experimental Situations

169   Our basic data consist of n pairs of observations, each pair containing one x-observation and one y-observation:

21.1
$$(x_1, y_1), \ldots, (x_n, y_n).$$

It is helpful to bring out the similarities and differences between the present setup and that of Chapter 11. In Chapter 11 we were concerned with two characteristics labeled A and B with sub-

198

categories $A_1, \ldots, A_c$ and $B_1, \ldots, B_r$. In the present case, in place of a finite number of subcategories of two characteristics A and B, we have two variables x and y measured along two continuous scales.

In Chapter 11 we considered two essentially different experimental setups. Under the first setup we selected n test subjects and classified each according to the appropriate combination $(A_j$ and $B_i)$. In the present case this corresponds to the determination of x- and y-values for n randomly selected test subjects. Under this setup both x- and y-values in (21.1) are subject to chance fluctuations.

Under the second setup discussed in Chapter 11 a single sample of size n is replaced by c samples of combined size n, one sample from each of the subcategories $A_1, \ldots, A_c$. In the present case this corresponds to the selection of c different x-values, say, $x_1, \ldots, x_c$, and the subsequent observation of the y-values of n test subjects, n_1 of whom are known to have x-value x_1, n_2 of whom are known to have x-value x_2, and so on, where $n_1 + \cdots + n_c = n$. Under this selection procedure only y-values are subject to chance fluctuations, while x-values are predetermined by the experimenter. The greater part of our discussion will be concerned with this second setup. Our n pairs of observations can then be written as

21.2
$$
\begin{array}{ccc}
(x_1, y_{11}) & \cdots & (x_c, y_{1c}) \\
(x_1, y_{21}) & & (x_c, y_{2c}) \\
\cdots & & \cdots \\
(x_1, y_{n_1 1}) & & (x_c, y_{n_c c})
\end{array}
$$

Observations of this type can be represented graphically by means of a *scatter diagram* by plotting the x-value in a given pair along a horizontal axis and the y-value along a vertical axis as in Figure 21.3.

FIGURE 21.3

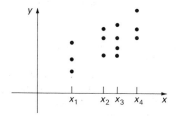

Scatter Diagram

Linear Regression

170 Experience shows that in many practical situations scatter diagrams exhibit a nearly linear trend. This suggests consideration of the following *linear regression* model: For every possible x-value we imagine a population of y-values. If we denote the mean of the y-population corresponding to a given value x by $\mu_{y.x}$, then $\mu_{y.x}$ considered as a function of x represents a straight line.*

Under the linearity assumption, we can write

21.4
$$\mu_{y.x} = \alpha' + \beta x$$

where α' and β are constants (not to be confused with type 1 and type 2 error probabilities). The expression in (21.4) is the simplest form for writing the equation of a straight line. But there are theoretical reasons for preferring a slightly different form. Let then

21.5
$$\bar{x} = \frac{n_1 x_1 + \cdots + n_c x_c}{n_1 + \cdots + n_c}$$

be the mean of the x-values selected for our experiment, each x-value weighted by the number of corresponding y-observations. (Alternatively $\bar{x}$ is the unweighted mean of all x's in (21.2).) If we then set $\alpha = \alpha' + \beta\bar{x}$, we can write (21.4) as

21.6
$$\mu_{y.x} = \alpha + \beta(x - \bar{x}).$$

The quantities α and β in (21.6) have the following statistical meaning: α is the mean of the y-population corresponding to $x = \bar{x}$; β equals the amount of increase (decrease if β is negative) in $\mu_{y.x}$ brought about by an increase of one unit in the value of x at which we observe y.

We can now write the following model for the observations y_{ji} in (21.2):

21.7
$$y_{ji} = \alpha + \beta(x_i - \bar{x}) + e_{ji}, \quad i = 1, \ldots, c, \quad j = 1, \ldots, n_i$$

where the e_{ji} are independently and identically distributed variables measuring by how much the y_{ji} deviate from their respective means $\mu_{y.x_i}$ due to random fluctuations. We shall usually assume that the e_{ji} are normal variables with mean 0 and the same (unknown) variance σ^2, $N(e_{ji}; 0, \sigma^2)$.

171   Model (21.7) allows us to state precisely the statistical problems that we want to solve:

(i) Find point estimates for the parameters α, β, $\mu_{y.x}$, and σ^2.
(ii) Find confidence intervals for the same parameters.
(iii) Test certain hypotheses about these parameters.

Of all possible hypotheses, the hypothesis

21.8
$$\beta = 0$$

is of greatest interest. If (21.8) is satisfied, we have

$$\mu_{y.x} = \alpha = \text{constant}$$

whatever the value x at which we decide to observe y. In other words, our y-variables do not depend on x.

---

*The subscript $y.x$ indicates that we are dealing with the mean of a y-population, where the y-population may depend on the value of some other measurement x.

172   The notation introduced in (21.2) is useful for establishing and understanding the basic model (21.7). However, for our subsequent discussions the double subscript notation is unnecessarily complicated. We note that if $n_1 = \cdots = n_c = 1$, (21.2) reduces to (21.1) with $n = c$. More generally, if we agree that in (21.1) the various x-values need not all be different, we can always use (21.1) to denote our observations. This we shall do from now on. In particular we shall often write our observations as (x_j, y_j), $j = 1, \ldots, n$. We note that in this simplified notation (21.5) reduces to $\bar{x} = \sum_j x_j/n$ and model (21.7) becomes

21.7′ $$y_j = \alpha + \beta(x_j - \bar{x}) + e_j, \quad N(e_j; 0, \sigma^2), \quad j = 1, \ldots, n.$$

The Principle of Least Squares

173   In the past when determining point estimates of population parameters, we usually appealed to intuition and common sense. When it comes to the estimation of the regression parameters α and β something more than intuition is required. What we need is a basic principle or method that tells us how to go about finding appropriate estimates. For the type of problem with which we are concerned, the *Principle of Least Squares* has been used by mathematicians for well over 100 years.

   In order to illustrate the Principle of Least Squares we return briefly to the problem of estimating the mean μ of a population on the basis of a random sample $x_1, \ldots, x_n$ taken from the population. According to the Principle of Least Squares the *least squares estimate m* of μ has the property that the sum of squared deviations of the n observations is smaller when taken from m than from any other number. More formally, the least squares estimate m minimizes the following sum of squares,

21.9 $$Q = \sum_j (y_j - m)^2.$$

In general, problems of minimization require the tools of calculus. But in the present case a simple argument allows us to find m directly. Let $\bar{x} = (x_1 + \cdots + x_n)/n$. We then have

$$Q = \sum_j [(x_j - \bar{x}) + (\bar{x} - m)]^2$$
$$= \sum_j (x_j - \bar{x})^2 + 2(\bar{x} - m) \sum_j (x_j - \bar{x}) + n(\bar{x} - m)^2$$
$$= \sum_j (x_j - \bar{x})^2 + n(\bar{x} - m)^2,$$

since $\sum_j (x_j - x) = 0$. The last expression is clearly minimized when m is taken to be $\bar{x}$. Thus the least squares estimate of μ is simply the sample mean $\bar{x}$ that we have used all along to estimate μ.

174   ***Point Estimates for*** $\mu_{y \cdot x}$, α ***and*** β We shall denote the point estimates of $\mu_{y \cdot x}$, α, and β by $m_{y \cdot x}$, a, and b, respectively. Since by (21.6), $\mu_{y \cdot x} = \alpha + \beta(x - \bar{x})$, we set

21.10 $$m_{y \cdot x} = a + b(x - \bar{x}),$$

where a and b are the least squares estimates of α and β.

For our regression problem, our observations are $y_1, \ldots, y_n$. In analogy to (21.9), the Principle of Least Squares requires that we minimize the following expression:

21.11
$$Q = \sum_j (y_j - m_{y \cdot x_j})^2 = \sum_j [y_j - a - b(x_j - \bar{x})]^2$$

in view of (21.10). By applying calculus methods (see Problem 12) it can be shown that the following expressions minimize Q:

21.12
$$a = \frac{1}{n} \sum_j y_j = \bar{y},$$

21.13
$$b = \frac{\sum_j (x_j - \bar{x})(y_j - \bar{y})}{\sum_j (x_j - \bar{x})^2}.$$

Then (21.10) becomes

21.14
$$m_{y \cdot x} = \bar{y} + b(x - \bar{x})$$

where b is given by (21.13). These are the least squares estimates of α, β, and $\mu_{y \cdot x}$.

175 The expressions in (21.12) and (21.13) can be obtained (more laboriously) by noncalculus methods as follows. We write

$$Q = \sum_j [(y_j - \bar{y}) - (a - \bar{y}) - b(x_j - \bar{x})]^2$$
$$= \sum_j (y_j - \bar{y})^2 + n(a - \bar{y})^2 + b^2 \sum_j (x_j - \bar{x})^2$$
$$- 2b \sum_j (x_j - \bar{x})(y_j - \bar{y}).$$

The other two terms involving crossproducts vanish since $\sum_j (x_j - \bar{x}) = \sum_j (y_j - \bar{y}) = 0$. Now

$$b^2 \sum_j (x_j - \bar{x})^2 - 2b \sum_j (x_j - \bar{x})(y_j - \bar{y})$$
$$= \sum_j (x_j - \bar{x})^2 \left[b - \frac{\sum_j (x_j - \bar{x})(y_j - \bar{y})}{\sum_j (x_j - \bar{x})^2} \right]^2$$
$$- \frac{[\sum_j (x_j - \bar{x})(y_j - \bar{y})]^2}{\sum_j (x_j - \bar{x})^2},$$

as is most easily seen by direct verification. Then

21.15
$$Q = \sum_j (y_j - \bar{y})^2 - \frac{[\sum_j (x_j - \bar{x})(y_j - \bar{y})]^2}{\sum_j (x_j - \bar{x})^2}$$
$$+ n(a - \bar{y})^2 + \sum_j (x_j - \bar{x})^2 \left[b - \frac{\sum_j (x_j - \bar{x})(y_j - \bar{y})}{\sum_j (x_j - \bar{x})^2} \right]^2.$$

Since the first two terms of Q do not involve a and b and the last two terms are non-negative, Q becomes a minimum when the last two terms are made as small as possible, namely zero. This occurs for a and b given by (21.12) and (21.13).

176 **Point Estimate for σ^2** In the case of a random sample $x_1, \ldots, x_n$ from a population with mean μ and variance σ^2, we estimate σ^2 as

$$s^2 = \frac{1}{n-1} \sum_j (x_j - \bar{x})^2,$$

$\bar{x}$ being the least squares estimate of the mean μ of the x_j. By analogy, in our regression model, we estimate the variance σ^2 occurring in (21.7') as

21.16
$$s^2 = \frac{1}{n-2} \sum_j (y_j - m_{y \cdot x_j})^2,$$

where, by (21.14), $m_{y \cdot x_j} = \bar{y} + b(x_j - \bar{x})$ is the least squares estimate of the mean $\mu_{y \cdot x_j} = \alpha + \beta(x_j - \bar{x})$.

177 **Computations** We use the following notation:

$$T_{xx} = \sum_j (x_j - \bar{x})^2 = \sum_j x_j^2 - \frac{1}{n} \left(\sum_j x_j \right)^2,$$

$$T_{yy} = \sum_j (y_j - \bar{y})^2 = \sum_j y_j^2 - \frac{1}{n} \left(\sum_j y_j \right)^2,$$

$$T_{xy} = \sum_j (x_j - \bar{x})(y_j - \bar{y}) = \sum_j x_j y_j - \bar{x} \sum_j y_j - \bar{y} \sum_j x_j + n \bar{x} \bar{y}$$

$$= \sum_j x_j y_j - \frac{1}{n} \left(\sum_j x_j \right) \left(\sum_j y_j \right).$$

As before,

21.12
$$a = \frac{1}{n} \sum_j y_j.$$

By (21.13), we compute b as

21.13'
$$b = \frac{T_{xy}}{T_{xx}}.$$

From (21.16), (21.11), and (21.15), we find that

21.17
$$(n-2)s^2 = \sum_j (y_j - m_{y \cdot x_j})^2 = Q$$
$$= \sum_j (y_j - \bar{y})^2 - \frac{[\sum_j (x_j - \bar{x})(y_j - \bar{y})]^2}{\sum_j (x_j - \bar{x})^2}$$
$$= T_{yy} - \frac{T_{xy}^2}{T_{xx}},$$

which, using (21.13'), we can also write as

21.18
$$(n-2)s^2 = \sum_j (y_j - m_{y \cdot x_j})^2$$
$$= T_{yy} - b^2 T_{xx}$$
$$= \sum_j (y_j - \bar{y})^2 - b^2 \sum_j (x_j - \bar{x})^2.$$

178 *An Example* In order to investigate how well grades on the midterm examination of a statistics course predict grades on the final examination, a professor selects 27 students according to their midterm grades and finds the following distribution of final examination grades:

TABLE
21.19

midterm grade (x)	45	55	65	75	85	95
number of students	1	4	6	6	6	4
final examination grade (y)	52	54	57	62	89	94
		63	72	77	93	97
		60	77	91	97	83
		62	80	71	74	95
			61	89	80	
			75	70	66	

Inspection of formulas (21.12), (21.13′), and (21.17) for a, b, and s^2 shows that we need the following information:

$$n = 27$$
$$\sum_j x_j = 45 + 4 \times 55 + \cdots + 4 \times 95 = 1995$$
$$\bar{x} = 73.9$$
$$\sum_j x_j^2 = 45^2 + 4 \times 55^2 + \cdots + 4 \times 95^2 = 152675$$
$$\sum_j y_j = 52 + 54 + \cdots + 95 = 2041$$
$$\bar{y} = 75.6$$
$$\sum_j y_j^2 = 52^2 + 54^2 + \cdots + 95^2 = 159427$$
$$\sum_j x_j y_j = 45 \times 52 + 55 \times 54 + \cdots + 95 \times 95 = 154885.$$

From these we find that

$$T_{xx} = \sum_j x_j^2 - (\sum_j x_j)^2/n = 5266.7$$
$$T_{yy} = \sum_j y_j^2 - (\sum_j y_j)^2/n = 5142.5$$
$$T_{xy} = \sum_j x_j y_j - (\sum_j x_j)(\sum_j y_j)/n = 4077.8.$$

It follows that our estimate of α is

$$\bar{y} = 75.6;$$

our estimate of β is

$$b = \frac{T_{xy}}{T_{xx}} = \frac{4077.8}{5266.7} = 0.774;$$

our estimate of σ^2 is

$$s^2 = \left(T_{yy} - \frac{T_{xy}^2}{T_{xx}}\right)/(n - 2)$$
$$= \left(5142.5 - \frac{4077.8^2}{5266.7}\right)/25$$
$$= 79.41;$$

and our estimate of σ is

$$s = \sqrt{79.41} = 8.91.$$

Finally we estimate $\mu_{y.x}$ as

21.20
$$\begin{aligned} m_{y.x} &= 75.6 + 0.774(x - 73.9) \\ &= 0.774x + 18.4. \end{aligned}$$

179 This last result can be used for prediction purposes. For students who have received 68 on the midterm examination, we predict a final examination grade of

$$0.774 \times 68 + 18.4 = 71.0.$$

For 88 on the midterm examination, we predict

$$0.774 \times 88 + 18.4 = 86.5$$

on the final examination.

Confidence Intervals and Tests of Hypotheses

180 We indicated in connection with our discussion of model (21.7) (or (21.7′)) that it is customary to assume that the deviations e_j, and therefore the observations y_j, are normal variables. Actually the least square approach for finding estimates for the regression parameters α and β does not depend on the normality assumption. However, in order to find confidence intervals and perform tests of hypotheses, we now specifically make the assumption that the e_j's are normal, $N(e_j; 0, \sigma^2)$. Using this assumption, the following three theorems are derived in courses of mathematical statistics:

THEOREM 21.21
The estimate $\bar{y}$ of α is normally distributed with mean α and variance σ^2/n.
(This is essentially the same theorem as the first part of Theorem 18.8.)

THEOREM 21.22
The estimate b of β is normally distributed with mean β and variance $\sigma^2/\sum_j (x_j - \bar{x})^2$.

THEOREM 21.23
The estimate $m_{y.x} = \bar{y} + b(x - \bar{x})$ of $\mu_{y.x} = \alpha + \beta(x - \bar{x})$ is normally distributed with mean $\mu_{y.x}$ and variance

$$\sigma^2 \left(\frac{1}{n} + \frac{(x - \bar{x})^2}{\sum_j (x_j - \bar{x})^2} \right).$$

These three theorems allow us to find confidence intervals and carry out tests of hypotheses by using the approach developed in Chapter 18. Since in regression problems the variance σ^2 is customarily unknown, we shall consider only this case.

181 Generalizing (18.15) we find the following three confidence intervals:

21.24
$$\bar{y} - t_\gamma s / \sqrt{n} \leq \alpha \leq \bar{y} + t_\gamma s / \sqrt{n};$$

21.25
$$b - t_\gamma s / \sqrt{\sum_j (x_j - \bar{x})^2} \leq \beta \leq b + t_\gamma s / \sqrt{\sum_j (x_j - \bar{x})^2};$$

21.26
$$m_{y \cdot x} - t_\gamma s \sqrt{\frac{1}{n} + \frac{(x - \bar{x})^2}{\sum_j (x_j - \bar{x})^2}}$$
$$\leq \mu_{y \cdot x} \leq m_{y \cdot x} + t_\gamma s \sqrt{\frac{1}{n} + \frac{(x - \bar{x})^2}{\sum_j (x_j - \bar{x})^2}}$$

where s is the square root of the expression in (21.16) and t_γ is read from Table H of the t-distribution with $n - 2$ degrees of freedom.

182 As usual, two-sided tests can be carried out by determining whether or not the hypothetical parameter value lies in the appropriate confidence interval. More directly, each of the following expressions has a t-distribution with $n - 2$ degrees of freedom:

21.27
$$t = \frac{\bar{y} - \alpha}{s / \sqrt{n}}$$

21.28
$$t = \frac{b - \beta}{s / \sqrt{\sum_j (x_j - \bar{x})^2}}$$

21.29
$$t = \frac{m_{y \cdot x} - \mu_{y \cdot x}}{s \sqrt{\frac{1}{n} + \frac{(x - \bar{x})^2}{\sum_j (x_j - \bar{x})^2}}} \cdot$$

We test a given hypothesis by setting the parameter in question equal to its hypothetical value and rejecting the hypothesis if the value of the resulting t-statistic differs significantly from zero.

As we pointed out the hypothesis (21.8) is often of particular interest, if $\beta = 0$, (21.28) reduces to

21.30
$$t = \frac{b \sqrt{\sum_j (x_j - \bar{x})^2}}{s},$$

and we reject the hypothesis that the y's do not depend on the x's when $|t|$, t, or $-t$ is sufficiently large, depending on whether under the alternative we have $\beta \neq 0$, $\beta > 0$, or $\beta < 0$.

183 It is instructive to develop the test (21.30) in a somewhat different form. For this purpose we rewrite (21.18) as

$$\sum_j (y_j - \bar{y})^2 = \sum_j (y_j - m_{y \cdot x_j})^2 + b^2 \sum_j (x_j - \bar{x})^2.$$

We can then construct the following analysis of variance table:

TABLE
21.31

source of variation	SS	df	MS
due to regression	$b^2 \sum_j (x_j - \bar{x})^2$	1	$b^2 \sum_j (x_j - \bar{x})^2$
about regression	$\sum_j (y_j - m_{y \cdot x_j})^2$	$n - 2$	$\sum_j (y_j - m_{y \cdot x_j})^2 / (n - 2) = s^2$
total	$\sum_j (y_j - \bar{y})^2$	$n - 1$	

We already know that s^2 is an estimate of σ^2, whether β is zero or not. On the other hand, it can be shown that $b^2 \sum_j (x_j - \bar{x})^2$ is an estimate of σ^2 only when $\beta = 0$. We can then test the hypothesis $\beta = 0$ by means of an F-test with

$$F = \frac{b^2 \sum_j (x_j - \bar{x})^2}{s^2}$$

and reject the hypothesis if F is sufficiently large according to the F-distribution with 1 numerator and $n - 2$ denominator degrees of freedom. It is easily seen from (21.30) that $F = t^2$. Rejecting the hypothesis $\beta = 0$ for large values of F corresponds to the two-sided t-test.

184 The terms "due to regression" and "about regression" in the analysis of variance table require some explanation. We can write

$$y_j - \bar{y} = (y_j - m_{y \cdot x_j}) + (m_{y \cdot x_j} - \bar{y}).$$

The term on the left measures by how much individual y-observations deviate from the mean $\bar{y}$ of all y-observations. The right side shows that this "total" variation is made up of two components: the variation of the individual y_j's *about* their regression estimates $m_{y \cdot x_j}$ and the deviation of these regression estimates from $\bar{y}$. The sum of squares "due to regression" measures the variation of the regression estimates caused by our choice of x-values. For this reason $\sum_j (m_{y \cdot x_j} - \bar{y})^2$ is also called *explained* variation, while $\sum_j (y_j - m_{y \cdot x_j})^2$ is called *unexplained* variation, unexplained because a knowledge of x does not explain by how much the actual observation y is going to deviate from its regression estimate $m_{y \cdot x}$.

185 We continue our analysis of the data (21.19). In particular we find confidence intervals with confidence coefficients $\gamma = .95$. With 25 degrees of freedom the critical t-value is 2.06. A confidence interval for α requires the computation of $t_\gamma s / \sqrt{n} = 2.06 \times 8.91 / \sqrt{27} = 3.53$. Then the confidence interval (21.24) becomes

$$75.6 - 3.5 \le \alpha \le 75.6 + 3.5$$

or

$$72.1 \le \alpha \le 79.1.$$

In order to find a confidence interval β, we need to find $t_\gamma s/\sqrt{\sum_j(x_j - \bar{x})^2} = 2.06 \times 8.91/\sqrt{5266.7} = 0.253$. Thus the confidence interval (21.25) becomes

$$0.774 - 0.253 \le \beta \le 0.774 + 0.253$$

or

$$0.521 \le \beta \le 1.027.$$

Finally let us again consider students with a midterm grade of 88 and find a confidence interval for their expected final examination scores. We have

$$t_\gamma s \sqrt{\frac{1}{n} + \frac{(x - \bar{x})^2}{\sum_j(x_j - \bar{x})^2}} = 2.06 \times 8.91 \sqrt{\frac{1}{27} + \frac{(88 - 73.9)^2}{5266.7}}$$
$$= 5.02,$$

and the confidence interval becomes

$$86.5 - 5.0 \le \mu_{y.88} \le 86.5 + 5.0$$

or

$$81.5 \le \mu_{y.88} \le 91.5.$$

It is important not to misinterpret this last result. We have found a confidence interval for $\mu_{y.88}$, the *average* final examination score of *all* students who obtain 88 on the midterm examination. A prediction interval within which a *particular* student's final grade will fall with, say, probability .95 would be considerably wider, since of course an individual student's score may deviate considerably from the group average.

186   The confidence interval for β already shows that the value $\beta = 0$ should be rejected. Final grades are not independent of midterm grades. But let us compute the appropriate t-statistic for a test of independence nevertheless. According to (21.30),

$$t = \frac{0.774\sqrt{5266.7}}{8.91} = 6.3.$$

Such a large value of t clearly implies that the hypothesis of independence should be rejected.

As an application of (21.29) let us consider the following problem. Suppose that it is claimed that students who receive a score of 90 on the midterm examination receive an average score of 95 on the final examination. We might try to refute such a claim as exaggerated by setting up the hypothesis $\mu_{y.90} = 95$ and testing it against the alternative $\mu_{y.90} < 95$. From (21.20),

$$m_{y.90} = 0.774 \times 90 + 18.4 = 88.1,$$

and therefore (21.29) becomes

$$t = \frac{88.1 - 95}{8.91 \sqrt{\frac{1}{27} + \frac{(90 - 73.9)^2}{5266.7}}} = -2.64.$$

For a one-sided test with significance level .05, the critical value is -1.71, so we reject our hypothesis.

The Correlation Coefficient r

187 In Chapter 16 we defined the Kendall rank correlation coefficient t_K as a measure of the strength of relationship between two rankings. (In Chapter 16 we wrote t, rather than t_K, but we shall use the subscripted symbol in this chapter to distinguish it from quantities that have a t-distribution.) While t_K can be computed for a set of n pairs of observations (x_j, y_j), in connection with linear regression it is customary to compute a different correlation coefficient, the *Pearson product moment* correlation coefficient denoted by the symbol r.

In order to define r, or rather r^2, we return to the analysis of variance representation (21.31) and set

$$r^2 = \frac{b^2 \sum_j (x_j - \bar{x})^2}{\sum_j (y_j - \bar{y})^2}.$$

According to this definition, r^2 equals the proportion of the total sum of squares explained by the regression of y on x. We then have

21.32

$$r = b \sqrt{\frac{\sum_j (x_j - \bar{x})^2}{\sum_j (y_j - \bar{y})^2}} = \frac{T_{xy}}{T_{xx}} \sqrt{\frac{T_{xx}}{T_{yy}}} = \frac{T_{xy}}{\sqrt{T_{xx} T_{yy}}}$$

$$= \frac{\sum_j (x_j - \bar{x})(y_j - \bar{y})}{\sqrt{\sum_j (x_j - \bar{x})^2 \sum_j (y_j - \bar{y})^2}}.$$

For the data (21.19) we find

$$r = \frac{T_{xy}}{\sqrt{T_{xx} T_{yy}}} = \frac{4077.8}{\sqrt{5266.7 \times 5142.5}} = .78.$$

Since $r^2 = .78^2 = .61$, we can say that 61% of the total variability among final examination grades can be explained in terms of the regression of final grades on midterm grades.

According to (21.32), r and b have the same sign: when b is positive, so is r; when b is negative, so is r; when b equals zero, so does r. As a consequence, a positive r implies that x and y tend to move in the same direction; a negative value of r implies that x and y tend to move in opposite directions.

188 We have seen that the Kendall rank correlation coefficient takes values between -1 and $+1$ and that $t_K = +1$ indicates identical rankings, while $t_K = -1$ indicates reverse rankings. We shall now show that similarly

$$-1 \le r \le +1,$$

and that $r = +1$ and $r = -1$ indicate complete agreement and complete disagreement between y's and x's in a sense to be specified now.

According to the analysis of variance representation we have

$$\sum_j (y_j - m_{y \cdot x_j})^2 = \sum_j (y_j - \bar{y})^2 - b^2 \sum_j (x_j - \bar{x})^2.$$

Since by the definition of r^2,

$$b^2 \sum_j (x_j - \bar{x})^2 = r^2 \sum_j (y_j - \bar{y})^2,$$

it follows that

$$\sum_j (y_j - m_{y \cdot x_j})^2 = (1 - r^2) \sum_j (y_j - \bar{y})^2.$$

When $r = \pm 1$, $r^2 = 1$ and $\sum_j (y_j - m_{y \cdot x_j})^2 = 0$. This can happen only when for every j, $y_j = m_{y \cdot x_j}$, that is, if all points (x_j, y_j) lie along the regression line $m_{y \cdot x} = \bar{y} + b(x - \bar{x})$. When $r = +1$, y-observations increase linearly together with x-observations; when $r = -1$, y-observations decrease linearly as x-observations increase as in Figure 21.33. Thus $r = +1$ implies $t_K = +1$ and $r = -1$ implies $t_K = -1$. But the reverse is not necessarily true. We can have $t_K = +1$ (or -1), without having $r = +1$ (or -1) (see Problem 10).

FIGURE 21.33

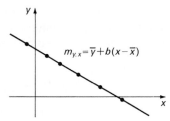

$$m_{y \cdot x} = \bar{y} + b(x - \bar{x})$$

189 Intuitively, a numerically small value of r suggests that y may not be dependent on x. More specifically, the test for the hypothesis of independence (21.30) can also be expressed in terms of the correlation coefficient r. Indeed,

$$t^2 = \frac{b^2 \sum_j (x_j - \bar{x})^2}{s^2} = \frac{(n - 2) b^2 \sum_j (x_j - \bar{x})^2}{\sum_j (y_j - m_{y \cdot x_j})^2} = \frac{(n - 2) r^2}{1 - r^2}$$

so that

21.34
$$t = \frac{r \sqrt{n - 2}}{\sqrt{1 - r^2}}.$$

It is easily seen that numerically large values of t suggesting rejection of the hypothesis of independence correspond to numerically large values of r. It is possible to compute r_γ-values corresponding to t_γ-values such that the hypothesis of independence is rejected at significance level $1 - \gamma$ whenever $r < -r_\gamma$ or $r > r_\gamma$.

Chance Fluctuations in x and y

190 So far we have assumed that the experimenter selects n values $x_1, \ldots, x_n$ (not necessarily all different) and then observes values

$y_1, \ldots, y_n$ corresponding to these x-values. Actually in practice the experimental setup is often different. Thus if x refers to a student's verbal score on the Graduate Record Examination (GRE) and y, to his quantitative score, we may select n students at random from the population of students taking the Graduate Record Examination and determine the n pairs of observations $(x_1, y_1), \ldots, (x_n, y_n)$. In such an experiment the x-values are not predetermined but are left to chance. Statisticians then may speak of a correlation model rather than a regression model, a distinction that is primarily historical and has little to recommend itself.

The earlier regression analysis remains valid even when x-values are not preselected but are determined by chance. However, in the latter case the correlation coefficient r also serves as an estimate of a parameter ρ, called the population correlation coefficient, that characterizes how much better one can predict y from x compared to not knowing x or, vice versa, how much better one can predict x from y compared to not knowing y. When analyzing data where both x's and y's are subject to chance it has been customary to concentrate primarily on ρ rather than on predicting one variable from the other.

When x's are selected by the experimenter, ρ is not defined. The experimenter's choice of x-values affects not only the accuracy with which the parameter β can be estimated but also the value of r. Theorem 21.22 shows that this accuracy is related to $\sum_j (x_j - \bar{x})^2$. The experimenter can increase his accuracy by choosing x's as far apart as possible. Of course practical and theoretical considerations, like the limited range of the linearity assumption, usually put rather definite bounds on how far x's can be spread out.

191   We have already mentioned that when both x- and y-variables are subject to chance we can predict not only y from x but also x from y. In the Graduate Record Examination example there is no reason to limit oneself to an investigation of how quantitative scores change with verbal scores. It is just as reasonable to ask how verbal scores change with quantitative scores.

We are then dealing with two different regression lines. In order to avoid confusion a precise notation is helpful. When talking of the regression of y on x we may write

$$\mu_{y \cdot x} = \alpha_{y \cdot x} + \beta_{y \cdot x}(x - \bar{x}),$$

which is estimated as

$$m_{y \cdot x} = \bar{y} + b_{y \cdot x}(x - \bar{x}),$$

where $b_{y \cdot x}$ (our former b) is given by (21.13). Correspondingly the regression of x on y is given by

$$\mu_{x \cdot y} = \alpha_{x \cdot y} + \beta_{x \cdot y}(y - \bar{y}),$$

which is estimated as

$$m_{x \cdot y} = \bar{x} + b_{x \cdot y}(y - \bar{y}),$$

where

$$b_{x \cdot y} = \frac{T_{xy}}{T_{yy}} = \frac{\sum_j (x_j - \bar{x})(y_j - \bar{y})}{\sum_j (y_j - \bar{y})^2}.$$

As far as the correlation coefficient r is concerned, (21.32) shows that r is symmetric in x and y and that therefore r remains unchanged under an interchange of x and y. Under the present setup, the test based on (21.34) becomes a test of the hypothesis $\rho = 0$, indicating that it does not pay to try predicting y from x or x from y.

General Regression Models

192   The regression problem discussed in Section 170 considers the simplest possible regression setup, where the mean of a variable y is a linear function of another variable x. We may try to generalize this situation by considering the regression of y on k variables $x_1, \ldots, x_k$. If y is the college grade point average of a student, x_1 may be his high school grade point average; x_2, his verbal Scholastic Aptitude Test score; x_3, his quantitative Scholastic Aptitude Test score. Clearly other variables can be added like IQ scores, class standing in high school, and so on. We then consider the following regression model for the mean of y-observations:

21.35
$$\mu_{y \cdot x_1 \ldots x_k} = \beta_0 + \beta_1 x_1 + \cdots + \beta_k x_k,$$

where $\beta_0, \beta_1, \ldots, \beta_k$ are unknown regression coefficients. The problem is to find estimates for and test hypotheses about these parameters.

While model (21.35) is linear in $x_1, \ldots, x_k$, it actually covers a much wider range. Thus if we set $x_1 = x$, $x_2 = x^2, \ldots, x_k = x^k$, we have the case of polynomial regression of y on a single variable x, where we assume that

21.36
$$\mu_{y \cdot x} = \beta_0 + \beta_1 x + \beta_2 x^2 + \cdots + \beta_k x^k.$$

By combining models (21.35) and (21.36) we can consider polynomial regression in more than one x-variable. The analysis of such problems is discussed in books dealing with advanced statistical methodology.

PROBLEMS   1 Assume you have computed the following information for 16 pairs of observations:

$$\sum_j x_j = 896 \qquad \sum_j y_j = 655$$
$$\sum_j x_j^2 = 52330 \qquad \sum_j y_j^2 = 29652$$
$$\sum_j x_j y_j = 38368.$$

Answer the following questions, making any appropriate assumptions.
   a. Find the regression of y on x.
   b. Estimate σ^2.

   c. Estimate $\mu_{y.50}$.

   d. Find a confidence interval for α.

   e. Find a confidence interval for β.

   f. Find a confidence interval for $\mu_{y.56}$ (use the same confidence coefficient as in part (d)).

   g. Compute r.

   h. What proportion of the total variability of y-observations can be explained in terms of the regression of y on x?

   i. Construct the analysis of variance table.

   j. Test for independence.

2 Compare the answers for parts (d) and (f) in Problem 1. Is this a general result or does it just happen to be true in this particular example?

3 It can be shown that for model (21.7′), $(n-2)s^2/\sigma^2 = \sum_j (y_j - m_{y.x})^2/\sigma^2$ has a chi-square distribution with $n-2$ degrees of freedom.

   a. Construct a confidence interval for σ^2. (*Hint:* Review Section 139.)

   b. Apply your solution under (a) to the data of Problem 1.

   c. Determine a test procedure for the hypothesis $\sigma^2 = \sigma_0^2$ against alternatives $\sigma^2 \neq \sigma_0^2$; $\sigma^2 < \sigma_0^2$; $\sigma^2 > \sigma_0^2$.

4 For Example 16.1:

   a. Construct a scatter diagram.

   b. Find the regression of psychology grades on statistics grades.

   c. Find the regression of statistics grades on psychology grades.

   d. Superimpose the two regression lines on the scatter diagram.

   e. Test for independence between psychology and statistics grades. What assumptions does your test imply?

5 For the data of Problem 5, Chapter 16:

   a. Find the Pearson product moment correlation coefficient.

   b. Test for independence.

6 Analyze the data of Problem 4, Chapter 12, by the methods of this chapter.

7 Compute r for the two rankings in Problem 3, Chapter 16. (This is called the *Spearman rank correlation coefficient*.)

8 For a sample of size 11, find r_γ such that we reject the hypothesis of independence at significance level $1 - \gamma = .01$ whenever $|r| > r_\gamma$.

9 Consider a scatter diagram with the least square regression line superimposed. What is the specific meaning of the quantity Q in (21.10)?

10 Find three pairs of numbers for which $|t_K| = 1$, but $|r| < 1$.

Problems 11 and 12 are for students with some knowledge of calculus.

11 Use calculus methods to show that (21.9) is a minimum for $m = \bar{x}$.

12 Let Q be given by (21.11). Show that

$$\frac{\partial Q}{\partial a} = -2\sum_j [y_j - a - b(x_j - \bar{x})]$$

$$\frac{\partial Q}{\partial b} = -2\sum_j [y_j - a - b(x_j - \bar{x})](x_j - \bar{x})$$

and that therefore the values of a and b that minimize (21.11) are given by (21.12) and (21.13).

Answers to Selected Problems

Chapter 1
1b   49.5
1d   39
3b   912, 1189

Chapter 3
1a $P(N) = .23$
1b $P(F \text{ and } N) = .075$
1c $P(L \text{ and } H) = .19$
1d $P(N \text{ or } SS) = .65$
1e $P(N \text{ or } S) = .385$
1f $P_s(N) = .225$
2a N and H, F and J
6a   .05, .4
9   .132

Chapter 4
1   2^n
7    $h(2, 0) = \frac{1}{36}, h(3, 1) = h(4, 2) = \frac{1}{8}$, etc.
13c npq
14b   1

Chapter 5
1    $b(2) = .069$
2    $b(2) = .195, b(8) = .003$
3   .421
4b   .299
4c   .213
6   .273
7a   .072
7b   .537

Chapter 6
1a   .9394
2a   .1335
3a   .0753
4a   .3954
5a   .58
6a   .524
7a   1.645

8a   .3085
9a   .885, .884
10   .949
12a   .964

Chapter 7
1   800
3   1024
5c   (.30, .50)
5d   (.32, .48)
9   no ($\gamma = .95$)
10   yes ($\gamma = .95$)

Chapter 8
1   .344
3a   .063
5   for Problem 9, $H_0: p = \frac{1}{3}$, rejected for $\alpha = .05$
6   accept

Chapter 9
3a   .089
3b   .578, .098, .007
5c   .382
7   .067
10a   approximately 210

Chapter 10
2a   test statistic $= 3.50$
4   yes
5   test statistic $= 21.9$

Chapter 11
3   test statistic $= 12.15$
6   test statistic $= .46$

Chapter 12
3    $S = 3, T = 7.5$
8a   .081
8b   .051

Chapter 13

1a accept

2b (14, 25), (14, 24.5) using
    $\gamma \doteq .96$

Chapter 14

4b $(-2, 23)$ using $\gamma = .947$

5a no $(\alpha \doteq .05)$

6   no $(\alpha \doteq .05)$

Chapter 15

1   test statistic $= 5.5$

2   test statistic $= 10.0$

3   test statistic $= 8.2$

6   test statistic $= 8.9$

Chapter 16

3a .78

3b test statistic $= 3.13$

7   yes (descriptive level $\doteq .013$)

Chapter 18

1a 77.1

2a 8.0

4a (74.9, 89.7)

4b test statistic $= 2.31$

4c approximately 106

6a claim does not seem justified

8b (2340, 4900) using $\gamma = .95$

Chapter 19

2b (14.5, 23.5) using $\gamma = .90$

4b $(-13.1, 3.1)$

8   test statistic $= 2.72$

Chapter 20

3a test statistic $= 2.05$

3d .95-confidence interval for
    $\psi = \mu_D - (\mu_A + \mu_B + \mu_C)/3$ is
    $(-3.3, 16.2)$

4a test statistic $= 22.45$

Chapter 21

1a $m_{y \cdot x} = 0.784x - 3.0$

1b 108.2

1c 36.2

1e (0.39, 1.18) using $\gamma = .90$

1g .68

1j test statistic $= 3.5$ or 12.2

3b (64.0, 230.6) using $\gamma = .90$

8   .735

Tables

Acknowledgments

The author is grateful for receiving permission to reproduce or adapt the following tables from the indicated sources:

Tables D, H, and I from *Biometrika Tables for Statisticians*, Volume I, Cambridge University Press, 1958.

Tables F and G from F. Wilcoxon, S. Katti, and R. A. Wilcox, *Critical Values and Probability Levels for the Wilcoxon Rank Sum Test and the Wilcoxon Signed Rank Test*, American Cyanamid Company, 1963, and Florida State University.

Tables J and K from *A Million Random Digits with 100,000 Normal Deviates*, The RAND Corporation, 1955.

TABLE A

Binomial Probabilities: $n = 10$

k \ p	.05	.10	.20	.30	.40	.50	
0	.599	.349	.107	.028	.006	.001	10
1	.315	.387	.268	.121	.040	.010	9
2	.075	.194	.302	.233	.121	.044	8
3	.010	.057	.201	.267	.215	.117	7
4	.001	.011	.088	.200	.251	.205	6
5		.001	.026	.103	.201	.246	5
6			.006	.037	.111	.205	4
7			.001	.009	.042	.117	3
8				.001	.011	.044	2
9					.002	.010	1
10						.001	0
	.95	.90	.80	.70	.60	.50	p \ k

Binomial Probabilities: $n = 13$

k \ p	.05	.10	.20	.30	$\frac{1}{3}$	.40	.50	
0	.513	.254	.055	.010	.005	.001		13
1	.351	.367	.179	.054	.034	.011	.002	12
2	.111	.245	.268	.139	.100	.045	.010	11
3	.021	.100	.246	.218	.183	.111	.035	10
4	.003	.028	.154	.234	.230	.184	.087	9
5		.006	.069	.180	.207	.221	.157	8
6		.001	.023	.103	.137	.197	.209	7
7			.006	.044	.069	.131	.209	6
8			.001	.014	.026	.066	.157	5
9				.003	.007	.024	.087	4
10				.001	.002	.006	.035	3
11						.001	.010	2
12							.002	1
13								0
	.95	.90	.80	.70	$\frac{2}{3}$	.60	.50	p \ k

TABLE A continued

Binomial Probabilities: $n = 20$

k \ p	.05	.10	.20	.30	.40	.50	
0	.358	.122	.012	.001			20
1	.377	.270	.058	.007			19
2	.189	.285	.137	.028	.003		18
3	.060	.190	.205	.072	.012	.001	17
4	.013	.090	.218	.130	.035	.005	16
5	.002	.032	.175	.179	.075	.015	15
6		.009	.109	.192	.124	.037	14
7		.002	.055	.164	.166	.074	13
8			.022	.114	.180	.120	12
9			.007	.065	.160	.160	11
10			.002	.031	.117	.176	10
11				.012	.071	.160	9
12				.004	.035	.120	8
13				.001	.015	.074	7
14					.005	.037	6
15					.001	.015	5
16						.005	4
17						.001	3
	.95	.90	.80	.70	.60	.50	p \ k

Binomial Probabilities: $n = 25$

p \ k	.05	.10	.20	.30	.40	.50	
0	.277	.072	.004				25
1	.365	.199	.024	.001			24
2	.231	.266	.071	.007			23
3	.093	.226	.136	.024	.002		22
4	.027	.138	.187	.057	.007		21
5	.006	.065	.196	.103	.020	.002	20
6	.001	.024	.163	.147	.044	.005	19
7		.007	.111	.171	.080	.014	18
8		.002	.062	.165	.120	.032	17
9			.029	.134	.151	.061	16
10			.012	.092	.161	.097	15
11			.004	.054	.147	.133	14
12			.001	.027	.114	.155	13
13				.011	.076	.155	12
14				.004	.043	.133	11
15				.001	.021	.097	10
16					.009	.061	9
17					.003	.032	8
18					.001	.014	7
19						.005	6
20						.002	5
	.95	.90	.80	.70	.60	.50	k / p

TABLE B

Normal Curve Areas: φ(z)

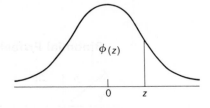

$\phi(z)$

z	.00	.01	.02	.03	.04	.05	.06	.07	.08	.09
.0	.5000	.5040	.5080	.5120	.5160	.5199	.5239	.5279	.5319	.5359
.1	.5398	.5438	.5478	.5517	.5557	.5596	.5636	.5675	.5714	.5753
.2	.5793	.5832	.5871	.5910	.5948	.5987	.6026	.6064	.6103	.6141
.3	.6179	.6217	.6255	.6293	.6331	.6368	.6406	.6443	.6480	.6517
.4	.6554	.6591	.6628	.6664	.6700	.6736	.6772	.6808	.6844	.6879
.5	.6915	.6950	.6985	.7019	.7054	.7088	.7123	.7157	.7190	.7224
.6	.7257	.7291	.7324	.7357	.7389	.7422	.7454	.7486	.7517	.7549
.7	.7580	.7611	.7642	.7673	.7704	.7734	.7764	.7794	.7823	.7852
.8	.7881	.7910	.7939	.7967	.7995	.8023	.8051	8078	.8106	.8133
.9	.8159	.8186	.8212	.8238	.8264	.8289	.8315	.8340	.8365	.8389
1.0	.8413	.8438	.8461	.8485	.8508	.8531	.8554	.8577	.8599	.8621
1.1	.8643	.8665	.8686	.8708	.8729	.8749	.8770	.8790	.8810	.8830
1.2	.8849	.8869	.8888	8907	.8925	.8944	.8962	.8980	.8997	.9015
1.3	.9032	.9049	.9066	.9082	.9099	.9115	.9131	.9147	.9162	.9177
1.4	.9192	.9207	.9222	.9236	.9251	.9265	.9279	.9292	.9306	.9319
1.5	.9332	.9345	.9357	.9370	.9382	.9394	.9406	.9418	.9429	.9441
1.6	.9452	.9463	.9474	.9484	.9495	.9505	.9515	.9525	.9535	.9545
1.7	.9554	.9564	.9573	.9582	.9591	.9599	.9608	.9616	.9625	.9633
1.8	.9641	.9649	.9656	.9664	.9671	.9678	.9686	.9693	.9699	.9706
1.9	.9713	.9719	.9726	.9732	.9738	.9744	.9750	.9756	.9761	.9767
2.0	.9772	.9778	.9783	.9788	.9793	.9798	.9803	.9808	.9812	.9817
2.1	.9821	.9826	.9830	.9834	.9838	.9842	.9846	.9850	.9854	.9857
2.2	.9861	.9864	.9868	.9871	.9875	.9878	.9881	.9884	.9887	.9890
2.3	.9893	.9896	.9898	.9901	.9904	.9906	.9909	.9911	.9913	.9916
2.4	.9918	.9920	.9922	.9925	.9927	.9929	.9931	.9932	.9934	.9936
2.5	.9938	.9940	.9941	.9943	.9945	.9946	.9948	.9949	.9951	.9952
2.6	.9953	.9955	.9956	.9957	.9959	.9960	.9961	.9962	.9963	.9964
2.7	.9965	.9966	.9967	.9968	.9969	.9970	.9971	.9972	.9973	.9974
2.8	.9974	.9975	.9976	.9977	.9977	.9978	.9979	.9979	.9980	.9981
2.9	.9981	.9982	.9982	.9983	.9984	.9984	.9985	.9985	.9986	.9986
3.0	.9987	.9987	.9987	.9988	.9988	.9989	.9989	.9989	.9990	.9990
3.1	.9990	.9991	.9991	.9991	.9992	.9992	.9992	.9992	.9993	.9993
3.2	.9993	.9993	.9994	.9994	.9994	.9994	.9994	.9995	.9995	.9995
3.3	.9995	.9995	.9995	.9996	.9996	.9996	.9996	.9996	.9996	.9997
3.4	.9997	.9997	.9997	.9997	.9997	.9997	.9997	.9997	.9997	.9998

TABLE C

Normal Deviations

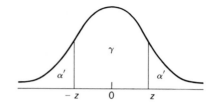

γ	$\alpha'' = 1 - \gamma$	$\alpha' = \frac{1}{2}(1 - \gamma)$	z
.995	.005	.0025	2.807
.99	.01	.005	2.576
.985	.015	.0075	2.432
.98	.02	.01	2.326
.975	.025	.0125	2.241
.97	.03	.015	2.170
.965	.035	.0175	2.108
.96	.04	.02	2.054
.954	.046	.023	2.000
.95	.05	.025	1.960
.94	.06	.03	1.881
.92	.08	.04	1.751
.9	.1	.05	1.645
.85	.15	.075	1.440
.8	.2	.10	1.282
.75	.25	.125	1.150
.7	.3	.150	1.036
.6	.4	.20	0.842
.5	.5	.25	0.674
.4	.6	.30	0.524
.3	.7	.35	0.385
.2	.8	.40	0.253
.1	.9	.45	0.126

γ = area between $-z$ and z
= confidence coefficient

$\alpha' = \frac{1}{2}(1 - \gamma)$
= area above z
= area below $-z$
= significance level for one-sided test

$\alpha'' = 1 - \gamma = 2\alpha'$
= area beyond $-z$ and z
= significance level for two-sided test

TABLE D

Chi-Square Distribution

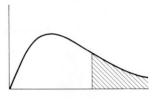

Upper Tail Probabilities

df	.990	.975	.950	.900	.750	.500	.250	.100	.050	.025	.010
1		.001	.004	.016	.102	.455	1.32	2.71	3.84	5.02	6.63
2	.020	.051	.103	.211	.575	1.39	2.77	4.61	5.99	7.38	9.21
3	.115	.216	.352	.584	1.21	2.37	4.11	6.25	7.81	9.35	11.3
4	.297	.484	.711	1.06	1.92	3.36	5.39	7.78	9.49	11.1	13.3
5	.554	.831	1.15	1.61	2.67	4.35	6.63	9.24	11.1	12.8	15.1
6	.872	1.24	1.64	2.20	3.45	5.35	7.84	10.6	12.6	14.4	16.8
7	1.24	1.69	2.17	2.83	4.25	6.35	9.04	12.0	14.1	16.0	18.5
8	1.65	2.18	2.73	3.49	5.07	7.34	10.2	13.4	15.5	17.5	20.1
9	2.09	2.70	3.33	4.17	5.90	8.34	11.4	14.7	16.9	19.0	21.7
10	2.56	3.25	3.94	4.87	6.74	9.34	12.5	16.0	18.3	20.5	23.2
11	3.05	3.82	4.57	5.58	7.58	10.3	13.7	17.3	19.7	21.9	24.7
12	3.57	4.40	5.23	6.30	8.44	11.3	14.8	18.5	21.0	23.3	26.2
13	4.11	5.01	5.89	7.04	9.30	12.3	16.0	19.8	22.4	24.7	27.7
14	4.66	5.63	6.57	7.79	10.2	13.3	17.1	21.1	23.7	26.1	29.1
15	5.23	6.26	7.26	8.55	11.0	14.3	18.2	22.3	25.0	27.5	30.6
16	5.81	6.91	7.96	9.31	11.9	15.3	19.4	23.5	26.3	28.8	32.0
17	6.41	7.56	8.67	10.1	12.8	16.3	20.5	24.8	27.6	30.2	33.4
18	7.01	8.23	9.39	10.9	13.7	17.3	21.6	26.0	28.9	31.5	34.8
19	7.63	8.91	10.1	11.7	14.6	18.3	22.7	27.2	30.1	32.9	36.2
20	8.26	9.59	10.9	12.4	15.5	19.3	23.8	28.4	31.4	34.2	37.6
21	8.90	10.3	11.6	13.2	16.3	20.3	24.9	29.6	32.7	35.5	38.9
22	9.54	11.0	12.3	14.0	17.2	21.3	26.0	30.8	33.9	36.8	40.3
23	10.2	11.7	13.1	14.8	18.1	22.3	27.1	32.0	35.2	38.1	41.6
24	10.9	12.4	13.8	15.7	19.0	23.3	28.2	33.2	36.4	39.4	43.0
25	11.5	13.1	14.6	16.5	19.9	24.3	29.3	34.4	37.7	40.6	44.3
26	12.2	13.8	15.4	17.3	20.8	25.3	30.4	35.6	38.9	41.9	45.6
27	12.9	14.6	16.2	18.1	21.7	26.3	31.5	36.7	40.1	43.2	47.0
28	13.6	15.3	16.9	18.9	22.7	27.3	32.6	37.9	41.3	44.5	48.3
29	14.3	16.0	17.7	19.8	23.6	28.3	33.7	39.1	42.6	45.7	49.6
30	15.0	16.8	18.5	20.6	24.5	29.3	34.8	40.3	43.8	47.0	50.9
40	22.2	24.4	26.5	29.1	33.7	49.3	45.6	51.8	55.8	59.3	63.7
60	37.5	40.5	43.2	46.5	52.3	59.3	67.0	74.4	79.1	83.3	88.4
	.010	.025	.050	.100	.250	.500	.750	.900	.950	.975	.990

Lower Tail Probabilities

For large *df*, the tabulated value equals approximately $\frac{1}{2}(z + \sqrt{2(df) - 1})^2$, where z is obtained from Table C corresponding to the appropriate upper or lower tail probability α'.

TABLE E

d-Factors for Sign Test and Confidence Intervals for the Median

γ = confidence coefficient
$\alpha' = \frac{1}{2}(1 - \gamma)$ = one-sided significance level
$\alpha'' = 2\alpha' = 1 - \gamma$ = two-sided significance level

n	d	γ	α''	α'	n	d	γ	α''	α'
3	1	.750	.250	.125	17	3	.998	.002	.001
4	1	.875	.125	.062		4	.987	.013	.006
5	1	.938	.062	.031		5	.951	.049	.025
6	1	.969	.031	.016		6	.857	.143	.072
	2	.781	.219	.109	18	4	.992	.008	.004
7	1	.984	.016	.008		5	.969	.031	.015
	2	.875	.125	.063		6	.904	.096	.048
8	1	.992	.008	.004		7	.762	.238	.119
	2	.930	.070	.035	19	4	.996	.004	.002
	3	.711	.289	.145		5	.981	.019	.010
9	1	.996	.004	.002		6	.936	.064	.032
	2	.961	.039	.020		7	.833	.167	.084
	3	.820	.180	.090	20	4	.997	.003	.001
10	1	.998	.002	.001		5	.988	.012	.006
	2	.979	.021	.011		6	.959	.041	.021
	3	.891	.109	.055		7	.885	.115	.058
11	1	.999	.001	.000	21	5	.993	.007	.004
	2	.988	.012	.006		6	.973	.027	.013
	3	.935	.065	.033		7	.922	.078	.039
	4	.773	.227	.113		8	.811	.189	.095
12	2	.994	.006	.003	22	5	.996	.004	.002
	3	.961	.039	.019		6	.983	.017	.008
	4	.854	.146	.073		7	.948	.052	.026
13	2	.997	.003	.002		8	.866	.134	.067
	3	.978	.022	.011	23	5	.997	.003	.001
	4	.908	.092	.046		6	.989	.011	.005
	5	.733	.267	.133		7	.965	.035	.017
14	2	.998	.002	.001		8	.907	.093	.047
	3	.987	.013	.006		9	.790	.210	.105
	4	.943	.057	.029	24	6	.993	.007	.003
	5_	.820	.180	.090		7	.977	.023	.011
15	3	.993	.007	.004		8	.936	.064	.032
	4	.965	.035	.018		9	.848	.152	.076
	5	.882	.118	.059	25	6	.996	.004	.002
16	3	.996	.004	.002		7	.985	.015	.007
	4	.979	.021	.011		8	.957	.043	.022
	5	.923	.077	.038		9	.892	.108	.054
	6	.790	.210	.105	26	7	.991	.009	.005

TABLE E continued

n	d	γ	α″	α′	n	d	γ	α″	α′
	8	.971	.029	.014	36	10	.996	.004	.002
	9	.924	.076	.038		11	.989	.011	.006
	10	.831	.169	.084		12	.971	.029	.014
27	7	.994	.006	.003		13	.935	.065	.033
	8	.981	.019	.010		14	.868	.132	.066
	9	.948	.052	.026	37	11	.992	.008	.004
	10	.878	.122	.061		12	.980	.020	.010
28	7	.996	.004	.002		13	.953	.047	.024
	8	.987	.013	.006		14	.901	.099	.049
	9	.964	.036	.018		15	.812	.188	.094
	10	.913	.087	.044	38	11	.995	.005	.003
	11	.815	.185	.092		12	.986	.014	.007
29	8	.992	.008	.004		13	.966	.034	.017
	9	.976	.024	.012		14	.927	.073	.036
	10	.939	.061	.031		15	.857	.143	.072
	11	.864	.136	.068	39	12	.991	.009	.005
30	8	.995	.005	.003		13	.976	.024	.012
	9	.984	.016	.008		14	.947	.053	.027
	10	.957	.043	.021		15	.892	.108	.054
	11	.901	.099	.049	40	12	.994	.006	.003
	12	.800	.200	.100		13	.983	.017	.008
31	8	.997	.003	.002		14	.962	.038	.019
	9	.989	.011	.005		15	.919	.081	.040
	10	.971	.029	.015		16	.846	.154	.077
	11	.929	.071	.035	41	12	.996	.004	.002
	12	.850	.150	.075		13	.988	.012	.006
32	9	.993	.007	.004		14	.972	.028	.014
	10	.980	.020	.010		15	.940	.060	.030
	11	.950	.050	.025		16	.883	.117	.059
	12	.890	.110	.055	42	13	.992	.008	.004
33	9	.995	.005	.002		14	.980	.020	.010
	10	.986	.014	.007		15	.956	.044	.022
	11	.965	.035	.018		16	.912	.088	.044
	12	.920	.080	.040		17	.836	.164	.082
	13	.837	.163	.081	43	13	.995	.005	.003
34	10	.991	.009	.005		14	.986	.014	.007
	11	.976	.024	.012		15	.968	.032	.016
	12	.942	.058	.029		16	.934	.066	.033
	13	.879	.121	.061		17	.874	.126	.063
35	10	.994	.006	.003	44	14	.990	.010	.005
	11	.983	.017	.008		15	.977	.023	.011
	12	.959	.041	.020		16	.951	.049	.024
	13	.910	.090	.045		17	.904	.096	.048
	14	.825	.175	.088		18	.826	.174	.087

n	d	γ	α″	α′	n	d	γ	α″	α′
45	14	.993	.007	.003	48	15	.994	.006	.003
	15	.984	.016	.008		16	.987	.013	.007
	16	.964	.036	.018		17	.971	.029	.015
	17	.928	.072	.036		18	.941	.059	.030
	18	.865	.135	.068		19	.889	.111	.056
46	14	.995	.005	.002	49	16	.991	.009	.005
	15	.989	.011	.006		17	.979	.021	.011
	16	.974	.026	.013		18	.956	.044	.022
	17	.946	.054	.027		19	.915	.085	.043
	18	.896	.104	.052		20	.848	.152	.076
47	15	.992	.008	.004	50	16	.993	.007	.003
	16	.981	.019	.009		17	.985	.015	.008
	17	.960	.040	.020		18	.967	033	.016
	18	.921	.079	.039		19	.935	.065	.032
	19	.856	.144	.072		20	.881	.119	.059

For $n > 50$ use $d \doteq \frac{1}{2}(n + 1 - z\sqrt{n})$, where z is read from Table C.

TABLE F

d-Factors for Wilcoxon Signed Rank Test and Confidence Intervals for the Median

γ = confidence coefficient
$\alpha' = \frac{1}{2}(1 - \gamma)$ = one-sided significance level
$\alpha'' = 2\alpha' = 1 - \gamma$ = two-sided significance level

n	d	γ	α''	α'	n	d	γ	α''	α'
3	1	.750	.250	.125	12	8	.991	.009	.005
4	1	.875	.125	.062		9	.988	.012	.006
5	1	.938	.062	.031		14	.958	.042	.021
	2	.875	.125	.063		15	.948	.052	.026
6	1	.969	.031	.016		18	.908	.092	.046
	2	.937	.063	.031		19	.890	.110	.055
	3	.906	.094	.047	13	10	.992	.008	.004
	4	.844	.156	.078		11	.990	.010	.005
7	1	.984	.016	.008		18	.952	.048	.024
	3	.953	.047	.016		19	.943	.057	.029
	4	.922	.078	.039		22	.906	.094	.047
	5	.891	.109	.055		23	.890	.110	.055
8	1	.992	.008	.004	14	13	.991	.009	.004
	2	.984	.016	.008		14	.989	.011	.005
	4	.961	.039	.020		22	.951	.049	.025
	5	.945	.055	.027		23	.942	.058	.029
	6	.922	.078	.039		26	.909	.091	.045
	7	.891	.109	.055		27	.896	.104	.052
9	2	.992	.008	.004	15	16	.992	.008	.004
	3	.988	.012	.006		17	.990	.010	.005
	6	.961	.039	.020		26	.952	.048	.024
	7	.945	.055	.027		27	.945	.055	.028
	9	.902	.098	.049		31	.905	.095	.047
	10	.871	.129	.065		32	.893	.107	.054
10	4	.990	.010	.005	16	20	.991	.009	.005
	5	.986	.014	.007		21	.989	.011	.006
	9	.951	.049	.024		30	.956	.044	.022
	10	.936	.064	.032		31	.949	.051	.025
	11	.916	.084	.042		36	.907	.093	.047
	12	.895	.105	.053		37	.895	.105	.052
11	6	.990	.010	.005	17	24	.991	.009	.005
	7	.986	.014	.007		25	.989	.011	.006
	11	.958	.042	.021		35	.955	.045	.022
	12	.946	.054	.027		36	.949	.051	.025
	14	.917	.083	.042		42	.902	.098	.049
	15	.898	.102	.051		43	.891	.109	.054

n	d	γ	α″	α′	n	d	γ	α″	α′
18	28	.991	.009	.005	22	49	.991	.009	.005
	29	.990	.010	.005		50	.990	.010	.005
	41	.952	.048	.024		66	.954	.046	.023
	42	.946	.054	.027		67	.950	.050	.025
	48	.901	.099	.049		76	.902	.098	.049
	49	.892	.108	.054		77	.895	.105	.053
19	33	.991	.009	.005	23	55	.991	.009	.005
	34	.989	.011	.005		56	.990	.010	.005
	47	.951	.049	.025		74	.952	.048	.024
	48	.945	.055	.027		75	.948	.052	.026
	54	.904	.096	.048		84	.902	.098	.049
	55	.896	.104	.052		85	.895	.105	.052
20	38	.991	.009	.005	24	62	.990	.010	.005
	39	.989	.011	.005		63	.989	.011	.005
	53	.952	.048	.024		82	.951	.049	.025
	54	.947	.053	.027		83	.947	.053	.026
	61	.903	.097	.049		92	.905	.095	.048
	62	.895	.105	.053		93	.899	.101	.051
21	43	.991	.009	.005	25	69	.990	.010	.005
	44	.990	.010	.005		70	.989	.011	.005
	59	.954	.046	.023		90	.952	.048	.024
	60	.950	.050	.025		91	.948	.052	.026
	68	.904	.096	.048		101	.904	.096	.048
	69	.897	.103	.052		102	.899	.101	.051

For $n > 25$ use $d \doteq \frac{1}{2}[\frac{1}{2}n(n + 1) + 1 - z\sqrt{n(n + 1)(2n + 1)/6}]$, where z is read from Table C.

TABLE G

d-Factors for Wilcoxon-Mann-Whitney Test and Confidence Intervals for the Shift Parameter Δ

γ = confidence coefficient

$\alpha' = \frac{1}{2}(1 - \gamma)$ = one-sided significance level

$\alpha'' = 2\alpha' = 1 - \gamma$ = two-sided significance level

		$m = 3$				$m = 4$		
	d	γ	α''	α'	d	γ	α''	α'
$n=3$	1	.900	.100	.050				
$n=4$	1	.943	.057	.029	1	.971	.029	.014
	2	.886	.114	.057	2	.943	.057	.029
					3	.886	.114	.057
$n=5$	1	.964	.036	.018	1	.984	.016	.008
	2	.929	.071	.036	2	.968	.032	.016
	3	.857	.143	.071	3	.937	.063	.032
					4	.889	.111	.056
$n=6$	1	.976	.024	.012	1	.990	.010	.005
	2	.952	.048	.024	2	.981	.019	.010
	3	.905	.095	.048	3	.962	.038	.019
	4	.833	.167	.083	4	.933	.067	.033
					5	.886	.114	.057
$n=7$	1	.983	.017	.008	1	.994	.006	.003
	2	.967	.033	.017	2	.988	.012	.006
	3	.933	.067	.033	4	.958	.042	.021
	4	.883	.117	.058	5	.927	.073	.036
					6	.891	.109	.055
$n=8$	1	.988	.012	.006	2	.992	.008	.004
	3	.952	.048	.024	3	.984	.016	.008
	4	.915	.085	.042	5	.952	.048	.024
	5	.867	.133	.067	6	.927	.073	.036
					7	.891	.109	.055
$n=9$	1	.991	.009	.005	2	.994	.006	.003
	2	.982	.018	.009	3	.989	.011	.006
	3	.964	.036	.018	5	.966	.034	.017
	4	.936	.064	.032	6	.950	.050	.025
	5	.900	.100	.050	7	.924	.076	.038
					8	.894	.106	.053
$n=10$	1	.993	.007	.004	3	.992	.008	.004
	2	.986	.014	.007	4	.986	.014	.007
	4	.951	.049	.025	6	.964	.036	.018
	5	.923	.077	.039	7	.946	.054	.027
	6	.888	.112	.056	8	.924	.076	.038
					9	.894	.106	.053
$n=11$	1	.995	.005	.003	3	.994	.006	.003
	2	.989	.011	.006	4	.990	.010	.005
	4	.962	.038	.019	7	.960	.040	.020
	5	.940	.060	.030	8	.944	.056	.028
	6	.912	.088	.044	9	.922	.078	.039
	7	.874	.126	.063	10	.896	.104	.052
$n=12$	2	.991	.009	.004	4	.992	.008	.004
	3	.982	.018	.009	5	.987	.013	.007
	5	.952	.048	.024	8	.958	.042	.021
	6	.930	.070	.035	9	.942	.058	.029
	7	.899	.101	.051	10	.922	.078	.039
					11	.897	.103	.052

For sample sizes m and n beyond the range of this table use $d \doteq \frac{1}{2}[mn + 1 - z\sqrt{mn(m + n + 1)/3}]$, where z is read from Table C.

		m = 5				m = 6				m = 7				m = 8		
	d	γ	α''	α'	d	γ	α''	α'	d	γ	α''	α'	d	γ	α''	α'
n = 5	1	.992	.008	.004												
	2	.984	.016	.008												
	3	.968	.032	.016												
	4	.944	.056	.028												
	5	.905	.095	.048												
	6	.849	.151	.075												
n = 6	2	.991	.009	.004	3	.991	.009	.004								
	3	.983	.017	.009	4	.985	.015	.008								
	4	.970	.030	.015	6	.959	.041	.021								
	5	.948	.052	.026	7	.935	.065	.033								
	6	.918	.082	.041	8	.907	.093	.047								
	7	.874	.126	.063	9	.868	.132	.066								
n = 7	2	.995	.005	.003	4	.992	.008	.004	5	.993	.007	.004				
	3	.990	.010	.005	5	.986	.014	.007	6	.989	.011	.006				
	6	.952	.048	.024	7	.965	.035	.018	9	.962	.038	.019				
	7	.927	.073	.037	8	.949	.051	.026	10	.947	.053	.027				
	8	.894	.106	.053	9	.927	.073	.037	12	.903	.097	.049				
					10	.899	.101	.051	13	.872	.128	.064				
n = 8	3	.994	.006	.003	5	.992	.008	.004	7	.991	.009	.005	8	.993	.007	.004
	4	.989	.011	.005	6	.987	.013	.006	8	.986	.014	.007	9	.990	.010	.005
	7	.955	.045	.023	9	.957	.043	.021	11	.960	.040	.020	14	.950	.050	.025
	8	.935	.065	.033	10	.941	.059	.030	12	.946	.054	.027	15	.935	.065	.033
	9	.907	.093	.047	11	.919	.081	.041	14	.906	.094	.047	16	.917	.083	.042
	10	.873	.127	.064	12	.892	.108	.054	15	.879	.121	.060	17	.895	.105	.052
n = 9	4	.993	.007	.004	6	.992	.008	.004	8	.992	.008	.004	10	.992	.008	.004
	5	.988	.012	.006	7	.988	.012	.006	9	.988	.012	.006	11	.989	.011	.006
	8	.958	.042	.021	11	.950	.050	.025	13	.958	.042	.021	16	.954	.046	.023
	9	.940	.060	.030	12	.934	.066	.033	14	.945	.055	.027	17	.941	.059	.030
	10	.917	.083	.042	13	.912	.088	.044	16	.909	.091	.045	19	.907	.093	.046
	11	.888	.112	.056	14	.887	.113	.057	17	.886	.114	.057	20	.886	.114	.057
n = 10	5	.992	.008	.004	7	.993	.007	.004	10	.990	.010	.005	12	.991	.009	.004
	6	.987	.013	.006	8	.989	.011	.006	11	.986	.014	.007	13	.988	.012	.006
	9	.960	.040	.020	12	.958	.042	.021	15	.957	.043	.022	18	.957	.043	.022
	10	.945	.055	.028	13	.944	.056	.028	16	.945	.055	.028	19	.945	.055	.027
	12	.901	.099	.050	15	.907	.093	.047	18	.912	.088	.044	21	.917	.083	.042
	13	.871	.129	.065	16	.882	.118	.059	19	.891	.109	.054	22	.899	.101	.051
n = 11	6	.991	.009	.004	8	.993	.007	.004	11	.992	.008	.004	14	.991	.009	.005
	7	.987	.013	.007	9	.990	.010	.005	12	.989	.011	.006	15	.988	.012	.006
	10	.962	.038	.019	14	.952	.048	.024	17	.956	.044	.022	20	.959	.041	.020
	11	.948	.052	.026	15	.938	.062	.031	18	.944	.056	.028	21	.949	.051	.025
	13	.910	.090	.045	17	.902	.098	.049	20	.915	.085	.043	24	.909	.091	.045
	14	.885	.115	.058	18	.878	.122	.061	21	.896	.104	.052	25	.891	.109	.054
n = 12	7	.991	.009	.005	10	.990	.010	.005	13	.990	.010	.005	16	.990	.010	.005
	8	.986	.014	.007	11	.987	.013	.007	14	.987	.013	.007	17	.988	.012	.006
	12	.952	.048	.024	15	.959	.041	.021	19	.955	.045	.023	23	.953	.047	.024
	13	.936	.064	.032	16	.947	.053	.026	20	.944	.056	.028	24	.943	.057	.029
	14	.918	.082	.041	18	.917	.083	.042	22	.917	.083	.042	27	.902	.098	.049
	15	.896	.104	.052	19	.898	.102	.051	23	.900	.100	.050	28	.885	.115	.058

TABLE G continued

	d	γ	α''	α'	d	γ	α''	α'	d	γ	α''	α'	d	γ	α''	α'
		m = 9				*m = 10*				*m = 11*				*m = 12*		
$n=9$	12	.992	.008	.004												
	13	.989	.011	.005												
	18	.960	.040	.020												
	19	.950	.050	.025												
	22	.906	.094	.047												
	23	.887	.113	.057												
$n=10$	14	.992	.008	.004	17	.991	.009	.005								
	15	.990	.010	.005	18	.989	.011	.006								
	21	.957	.043	.022	24	.957	.043	.022								
	22	.947	.053	.027	25	.948	.052	.026								
	25	.905	.095	.047	28	.911	.089	.045								
	26	.887	.113	.056	29	.895	.105	.053								
$n=11$	17	.990	.010	.005	19	.992	.008	.004	22	.992	.008	.004				
	18	.988	.012	.006	20	.990	.010	.005	23	.989	.011	.005				
	24	.954	.046	.023	27	.957	.043	.022	31	.953	.047	.024				
	25	.944	.056	.028	28	.949	.051	.026	32	.944	.056	.028				
	28	.905	.095	.048	32	.901	.099	.049	35	.912	.088	.044				
	29	.888	.112	.056	33	.886	.114	.057	36	.899	.101	.051				
$n=12$	19	.991	.009	.005	22	.991	.009	.005	25	.991	.009	.004	28	.992	.008	.004
	20	.988	.012	.006	23	.989	.011	.006	26	.989	.011	.005	29	.990	.010	.005
	27	.951	.049	.025	30	.957	.043	.021	34	.956	.044	.022	38	.955	.045	.023
	28	.942	.058	.029	31	.950	.050	.025	35	.949	.051	.026	39	.948	.052	.026
	31	.905	.095	.048	35	.907	.093	.047	39	.909	.091	.045	43	.911	.089	.044
	32	.889	.111	.056	36	.893	.107	.054	40	.896	.104	.052	44	.899	.101	.050

TABLE H

t-Distribution

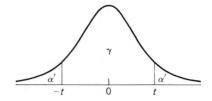

γ	.5	.8	.9	.95	.98	.99
α'	.25	.1	.05	.025	.01	.005
α''	.5	.2	.1	.05	.02	.01
df						
1	1.000	3.078	6.314	12.706	31.821	63.657
2	.816	1.886	2.920	4.303	6.965	9.925
3	.765	1.638	2.353	3.182	4.541	5.841
4	.741	1.533	2.132	2.776	3.747	4.604
5	.727	1.476	2.015	2.571	3.365	4.032
6	.718	1.440	1.943	2.447	3.143	3.707
7	.711	1.415	1.895	2.365	2.998	3.499
8	.706	1.397	1.860	2.306	2.896	3.355
9	.703	1.383	1.833	2.262	2.821	3.250
10	.700	1.372	1.812	2.228	2.764	3.169
11	.697	1.363	1.796	2.201	2.718	3.106
12	.695	1.356	1.782	2.179	2.681	3.055
13	.694	1.350	1.771	2.160	2.650	3.012
14	.692	1.345	1.761	2.145	2.624	2.977
15	.691	1.341	1.753	2.131	2.602	2.947
16	.690	1.337	1.746	2.120	2.583	2.921
17	.689	1.333	1.740	2.110	2.567	2.898
18	.688	1.330	1.734	2.101	2.552	2.878
19	.688	1.328	1.729	2.093	2.539	2.861
20	.687	1.325	1.725	2.086	2.528	2.845
21	.686	1.323	1.721	2.080	2.518	2.831
22	.686	1.321	1.717	2.074	2.508	2.819
23	.685	1.319	1.714	2.069	2.500	2.807
24	.685	1.318	1.711	2.064	2.492	2.797
25	.684	1.316	1.708	2.060	2.485	2.787
26	.684	1.315	1.706	2.056	2.479	2.779
27	.684	1.314	1.703	2.052	2.473	2.771
28	.683	1.313	1.701	2.048	2.467	2.763
29	.683	1.311	1.699	2.045	2.462	2.756
30	.683	1.310	1.697	2.042	2.457	2.750
40	.681	1.303	1.684	2.021	2.423	2.704
60	.679	1.296	1.671	2.000	2.390	2.660
120	.677	1.289	1.658	1.980	2.358	2.617
∞	.674	1.282	1.645	1.960	2.326	2.576

γ = area between $-t$ and t
= confidence coefficient

$\alpha' = \frac{1}{2}(1 - \gamma)$
= area above t
= area below $-t$
= significance level of one-sided test

$\alpha'' = 1 - \gamma = 2\alpha'$
= area beyond $-t$ and t
= significance level of two-sided test

TABLE I

F-Distribution

Upper tail probability $\alpha = .10$

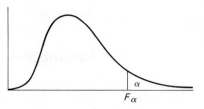

Numerator Degrees of Freedom

		1	2	3	4	5	6	7	8	9
	1	39.86	49.50	53.59	55.83	57.24	58.20	58.91	59.44	59.86
	2	8.53	9.00	9.16	9.24	9.29	9.33	9.35	9.37	9.38
	3	5.54	5.46	5.39	5.34	5.31	5.28	5.27	5.25	5.24
	4	4.54	4.32	4.19	4.11	4.05	4.01	3.98	3.95	3.94
	5	4.06	3.78	3.62	3.52	3.45	3.40	3.37	3.34	3.32
	6	3.78	3.46	3.29	3.18	3.11	3.05	3.01	2.98	2.96
	7	3.59	3.26	3.07	2.96	2.88	2.83	2.78	2.75	2.72
	8	3.46	3.11	2.92	2.81	2.73	2.67	2.62	2.59	2.56
	9	3.36	3.01	2.81	2.69	2.61	2.55	2.51	2.47	2.44
Denominator Degrees of Freedom	10	3.29	2.92	2.73	2.61	2.52	2.46	2.41	2.38	2.35
	11	3.23	2.86	2.66	2.54	2.45	2.39	2.34	2.30	2.27
	12	3.18	2.81	2.61	2.48	2.39	2.33	2.28	2.24	2.21
	13	3.14	2.76	2.56	2.43	2.35	2.28	2.23	2.20	2.16
	14	3.10	2.73	2.52	2.39	2.31	2.24	2.19	2.15	2.12
	15	3.07	2.70	2.49	2.36	2.27	2.21	2.16	2.12	2.09
	16	3.05	2.67	2.46	2.33	2.24	2.18	2.13	2.09	2.06
	17	3.03	2.64	2.44	2.31	2.22	2.15	2.10	2.06	2.03
	18	3.01	2.62	2.42	2.29	2.20	2.13	2.08	2.04	2.00
	19	2.99	2.61	2.40	2.27	2.18	2.11	2.06	2.02	1.98
	20	2.97	2.59	2.38	2.25	2.16	2.09	2.04	2.00	1.96
	21	2.96	2.57	2.36	2.23	2.14	2.08	2.02	1.98	1.95
	22	2.95	2.56	2.35	2.22	2.13	2.06	2.01	1.97	1.93
	23	2.94	2.55	2.34	2.21	2.11	2.05	1.99	1.95	1.92
	24	2.93	2.54	2.33	2.19	2.10	2.04	1.98	1.94	1.91
	25	2.92	2.53	2.32	2.18	2.09	2.02	1.97	1.93	1.89
	26	2.91	2.52	2.31	2.17	2.08	2.01	1.96	1.92	1.88
	27	2.90	2.51	2.30	2.17	2.07	2.00	1.95	1.91	1.87
	28	2.89	2.50	2.29	2.16	2.06	2.00	1.94	1.90	1.87
	29	2.89	2.50	2.28	2.15	2.06	1.99	1.93	1.89	1.86
	30	2.88	2.49	2.28	2.14	2.05	1.98	1.93	1.88	1.85
	40	2.84	2.44	2.23	2.09	2.00	1.93	1.87	1.83	1.79
	60	2.79	2.39	2.18	2.04	1.95	1.87	1.82	1.77	1.74
	120	2.75	2.35	2.13	1.99	1.90	1.82	1.77	1.72	1.68
	∞	2.71	2.30	2.08	1.94	1.85	1.77	1.72	1.67	1.63

Numerator Degrees of Freedom

10	12	15	20	24	30	40	60	120	∞
60.19	60.71	61.22	61.74	62.00	62.26	62.53	62.79	63.06	63.33
9.39	9.41	9.42	9.44	9.45	9.46	9.47	9.47	9.48	9.49
5.23	5.22	5.20	5.18	5.18	5.17	5.16	5.15	5.14	5.13
3.92	3.90	3.87	3.84	3.83	3.82	3.80	3.79	3.78	3.76
3.30	3.27	3.24	3.21	3.19	3.17	3.16	3.14	3.12	3.10
2.94	2.90	2.87	2.84	2.82	2.80	2.78	2.76	2.74	2.72
2.70	2.67	2.63	2.59	2.58	2.56	2.54	2.51	2.49	2.47
2.54	2.50	2.46	2.42	2.40	2.38	2.36	2.34	2.32	2.29
2.42	2.38	2.34	2.30	2.28	2.25	2.23	2.21	2.18	2.16
2.32	2.28	2.24	2.20	2.18	2.16	2.13	2.11	2.08	2.06
2.25	2.21	2.17	2.12	2.10	2.08	2.05	2.03	2.00	1.97
2.19	2.15	2.10	2.06	2.04	2.01	1.99	1.96	1.93	1.90
2.14	2.10	2.05	2.01	1.98	1.96	1.93	1.90	1.88	1.85
2.10	2.05	2.01	1.96	1.94	1.91	1.89	1.86	1.83	1.80
2.06	2.02	1.97	1.92	1.90	1.87	1.85	1.82	1.79	1.76
2.03	1.99	1.94	1.89	1.87	1.84	1.81	1.78	1.75	1.72
2.00	1.96	1.91	1.86	1.84	1.81	1.78	1.75	1.72	1.69
1.98	1.93	1.89	1.84	1.81	1.78	1.75	1.72	1.69	1.66
1.96	1.91	1.86	1.81	1.79	1.76	1.73	1.70	1.67	1.63
1.94	1.89	1.84	1.79	1.77	1.74	1.71	1.68	1.64	1.61
1.92	1.87	1.83	1.78	1.75	1.72	1.69	1.66	1.62	1.59
1.90	1.86	1.81	1.76	1.73	1.70	1.67	1.64	1.60	1.57
1.89	1.84	1.80	1.74	1.72	1.69	1.66	1.62	1.59	1.55
1.88	1.83	1.78	1.73	1.70	1.67	1.64	1.61	1.57	1.53
1.87	1.82	1.77	1.72	1.69	1.66	1.63	1.59	1.56	1.52
1.86	1.81	1.76	1.71	1.68	1.65	1.61	1.58	1.54	1.50
1.85	1.80	1.75	1.70	1.67	1.64	1.60	1.57	1.53	1.49
1.84	1.79	1.74	1.69	1.66	1.63	1.59	1.56	1.52	1.48
1.83	1.78	1.73	1.68	1.65	1.62	1.58	1.55	1.51	1.47
1.82	1.77	1.72	1.67	1.64	1.61	1.57	1.54	1.50	1.46
1.76	1.71	1.66	1.61	1.57	1.54	1.51	1.47	1.42	1.38
1.71	1.66	1.60	1.54	1.51	1.48	1.44	1.40	1.35	1.29
1.65	1.60	1.55	1.48	1.45	1.41	1.37	1.32	1.26	1.19
1.60	1.55	1.49	1.42	1.38	1.34	1.30	1.24	1.17	1.00

TABLE I continued

Upper tail probability $\alpha = .05$

Numerator Degrees of Freedom

		1	2	3	4	5	6	7	8	9
	1	161.4	199.5	215.7	224.6	230.2	234.0	236.8	238.9	240.5
	2	18.15	19.00	19.16	19.25	19.30	19.33	19.35	19.37	19.38
	3	10.13	9.55	9.28	9.12	9.01	8.94	8.89	8.85	8.81
	4	7.71	6.94	6.59	6.39	6.26	6.16	6.09	6.04	6.00
	5	6.61	5.79	5.41	5.19	5.05	4.95	4.88	4.82	4.77
	6	5.99	5.14	4.76	4.53	4.39	4.28	4.21	4.15	4.10
	7	5.59	4.74	4.35	4.12	3.97	3.87	3.79	3.73	3.68
	8	5.32	4.46	4.07	3.84	3.69	3.58	3.50	3.44	3.39
	9	5.12	4.26	3.86	3.63	3.48	3.37	3.29	3.23	3.18
	10	4.96	4.10	3.71	3.48	3.33	3.22	3.14	3.07	3.02
	11	4.84	3.98	3.59	3.36	3.20	3.09	3.01	2.95	2.90
	12	4.75	3.89	3.49	3.26	3.11	3.00	2.91	2.85	2.80
Denominator Degrees of Freedom	13	4.67	3.81	3.41	3.18	3.03	2.92	2.83	2.77	2.71
	14	4.60	3.74	3.34	3.11	2.96	2.85	2.76	2.70	2.65
	15	4.54	3.68	3.29	3.06	2.90	2.79	2.71	2.64	2.59
	16	4.49	3.63	3.24	3.01	2.85	2.74	2.66	2.59	2.54
	17	4.45	3.59	3.20	2.96	2.81	2.70	2.61	2.55	2.49
	18	4.41	3.55	3.16	2.93	2.77	2.66	2.58	2.51	2.46
	19	4.38	3.52	3.13	2.90	2.74	2.63	2.54	2.48	2.42
	20	4.35	3.49	3.10	2.87	2.71	2.60	2.51	2.45	2.39
	21	4.32	3.47	3.07	2.84	2.68	2.57	2.49	2.42	2.37
	22	4.30	3.44	3.05	2.82	2.66	2.55	2.46	2.40	2.34
	23	4.28	3.42	3.03	2.80	2.64	2.53	2.44	2.37	2.32
	24	4.26	3.40	3.01	2.78	2.62	2.51	2.42	2.36	2.30
	25	4.24	3.39	2.99	2.76	2.60	2.49	2.40	2.34	2.28
	26	4.23	3.37	2.98	2.74	2.59	2.47	2.39	2.32	2.27
	27	4.21	3.35	2.96	2.73	2.57	2.46	2.37	2.31	2.25
	28	4.20	3.34	2.95	2.71	2.56	2.45	2.36	2.29	2.24
	29	4.18	3.33	2.93	2.70	2.55	2.43	2.35	2.28	2.22
	30	4.17	3.32	2.92	2.69	2.53	2.42	2.33	2.27	2.21
	40	4.08	3.23	2.84	2.61	2.45	2.34	2.25	2.18	2.12
	60	4.00	3.15	2.76	2.53	2.37	2.25	2.17	2.10	2.04
	120	3.92	3.07	2.68	2.45	2.29	2.17	2.09	2.02	1.96
	∞	3.84	3.00	2.60	2.37	2.21	2.10	2.01	1.94	1.88

Numerator Degrees of Freedom

10	12	15	20	24	30	40	60	120	∞
241.9	243.9	245.9	248.0	249.1	250.1	251.1	252.2	253.3	254.3
19.40	19.41	19.43	19.45	19.45	19.46	19.47	19.48	19.49	19.50
8.79	8.74	8.70	8.66	8.64	8.62	8.59	8.57	8.55	8.53
5.96	5.91	5.86	5.80	5.77	5.75	5.72	5.69	5.66	5.63
4.74	4.68	4.62	4.56	4.53	4.50	4.46	4.43	4.40	4.36
4.06	4.00	3.94	3.87	3.84	3.81	3.77	3.74	3.70	3.67
3.64	3.57	3.51	3.44	3.41	3.38	3.34	3.30	3.27	3.23
3.35	3.28	3.22	3.15	3.12	3.08	3.04	3.01	2.97	2.93
3.14	3.07	3.01	2.94	2.90	2.86	2.83	2.79	2.75	2.71
2.98	2.91	2.85	2.77	2.74	2.70	2.66	2.62	2.58	2.54
2.85	2.79	2.72	2.65	2.61	2.57	2.53	2.49	2.45	2.40
2.75	2.69	2.62	2.54	2.51	2.47	2.43	2.38	2.34	2.30
2.67	2.60	2.53	2.46	2.42	2.38	2.34	2.30	2.25	2.21
2.60	2.53	2.46	2.39	2.35	2.31	2.27	2.22	2.18	2.13
2.54	2.48	2.40	2.33	2.29	2.25	2.20	2.16	2.11	2.07
2.49	2.42	2.35	2.28	2.24	2.19	2.15	2.11	2.06	2.01
2.45	2.38	2.31	2.23	2.19	2.15	2.10	2.06	2.01	1.96
2.41	2.34	2.27	2.19	2.15	2.11	2.06	2.02	1.97	1.92
2.38	2.31	2.23	2.16	2.11	2.07	2.03	1.98	1.93	1.88
2.35	2.28	2.20	2.12	2.08	2.04	1.99	1.95	1.90	1.84
2.32	2.25	2.18	2.10	2.05	2.01	1.96	1.92	1.87	1.81
2.30	2.23	2.15	2.07	2.03	1.98	1.94	1.89	1.84	1.78
2.27	2.20	2.13	2.05	2.01	1.96	1.91	1.86	1.81	1.76
2.25	2.18	2.11	2.03	1.98	1.94	1.89	1.84	1.79	1.73
2.24	2.16	2.09	2.01	1.96	1.92	1.87	1.82	1.77	1.71
2.22	2.15	2.07	1.99	1.95	1.90	1.85	1.80	1.75	1.69
2.20	2.13	2.06	1.97	1.93	1.88	1.84	1.79	1.73	1.67
2.19	2.12	2.04	1.96	1.91	1.87	1.82	1.77	1.71	1.65
2.18	2.10	2.03	1.94	1.90	1.85	1.81	1.75	1.70	1.64
2.16	2.09	2.01	1.93	1.89	1.84	1.79	1.74	1.68	1.62
2.08	2.00	1.92	1.84	1.79	1.74	1.69	1.64	1.58	1.51
1.99	1.92	1.84	1.75	1.70	1.65	1.59	1.53	1.47	1.39
1.91	1.83	1.75	1.66	1.61	1.55	1.50	1.43	1.35	1.25
1.83	1.75	1.67	1.57	1.52	1.46	1.39	1.32	1.22	1.00

TABLE I continued

Upper tail probability $\alpha = .025$

Numerator Degrees of Freedom

	1	2	3	4	5	6	7	8	9
1	647.8	799.5	864.2	899.6	921.8	937.1	948.2	956.7	963.3
2	38.51	39.00	39.17	39.25	39.30	39.33	39.36	39.37	39.39
3	17.44	16.04	15.44	15.10	14.88	14.73	14.62	14.54	14.47
4	12.22	10.65	9.98	9.60	9.36	9.20	9.07	8.98	8.90
5	10.01	8.43	7.76	7.39	7.15	6.98	6.85	6.76	6.68
6	8.81	7.26	6.60	6.23	5.99	5.82	5.70	5.60	5.52
7	8.07	6.54	5.89	5.52	5.29	5.12	4.99	4.90	4.82
8	7.57	6.06	5.42	5.05	4.82	4.65	4.53	4.43	4.36
9	7.21	5.71	5.08	4.72	4.48	4.32	4.20	4.10	4.03
10	6.94	5.46	4.83	4.47	4.24	4.07	3.95	3.85	3.78
11	6.72	5.26	4.63	4.28	4.04	3.88	3.76	3.66	3.59
12	6.55	5.10	4.47	4.12	3.89	3.73	3.61	3.51	3.44
13	6.41	4.97	4.35	4.00	3.77	3.60	3.48	3.39	3.31
14	6.30	4.86	4.24	3.89	3.66	3.50	3.38	3.29	3.21
15	6.20	4.77	4.15	3.80	3.58	3.41	3.29	3.20	3.12
16	6.12	4.69	4.08	3.73	3.50	3.34	3.22	3.12	3.05
17	6.04	4.62	4.01	3.66	3.44	3.28	3.16	3.06	2.98
18	5.98	4.56	3.95	3.61	3.38	3.22	3.10	3.01	2.93
19	5.92	4.51	3.90	3.56	3.33	3.17	3.05	2.96	2.88
20	5.87	4.46	3.86	3.51	3.29	3.13	3.01	2.91	2.84
21	5.83	4.42	3.82	3.48	3.25	3.09	2.97	2.87	2.80
22	5.79	4.38	3.78	3.44	3.22	3.05	2.93	2.84	2.76
23	5.75	4.35	3.75	3.41	3.18	3.02	2.90	2.81	2.73
24	5.72	4.32	3.72	3.38	3.15	2.99	2.87	2.78	2.70
25	5.69	4.29	3.69	3.35	3.13	2.97	2.85	2.75	2.68
26	5.66	4.27	3.67	3.33	3.10	2.94	2.82	2.73	2.65
27	5.63	4.24	3.65	3.31	3.08	2.92	2.80	2.71	2.63
28	5.61	4.22	3.63	3.29	3.06	2.90	2.78	2.69	2.61
29	5.59	4.20	3.61	3.27	3.04	2.88	2.76	2.67	2.59
30	5.57	4.18	3.59	3.25	3.03	2.87	2.75	2.65	2.57
40	5.42	4.05	3.46	3.13	2.90	2.74	2.62	2.53	2.45
60	5.29	3.93	3.34	3.01	2.79	2.63	2.51	2.41	2.33
120	5.15	3.80	3.23	2.89	2.67	2.52	2.39	2.30	2.22
∞	5.02	3.69	3.12	2.79	2.57	2.41	2.29	2.19	2.11

Denominator Degrees of Freedom

Numerator Degrees of Freedom

10	12	15	20	24	30	40	60	120	∞
968.6	976.7	984.9	993.1	997.2	1001	1006	1010	1014	1018
39.40	39.41	39.43	39.45	39.46	39.46	39.47	39.48	39.49	39.50
14.42	14.34	14.25	14.17	14.12	14.08	14.04	13.99	13.95	13.90
8.84	8.75	8.66	8.56	8.51	8.46	8.41	8.36	8.31	8.26
6.62	6.52	6.43	6.33	6.28	6.23	6.18	6.12	6.07	6.02
5.46	5.37	5.27	5.17	5.12	5.07	5.01	4.96	4.90	4.85
4.76	4.67	4.57	4.47	4.42	4.36	4.31	4.25	4.20	4.14
4.30	4.20	4.10	4.00	3.95	3.89	3.84	3.78	3.73	3.67
3.96	3.87	3.77	3.67	3.61	3.56	3.51	3.45	3.39	3.33
3.72	3.62	3.52	3.42	3.37	3.31	3.26	3.20	3.14	3.08
3.53	3.43	3.33	3.23	3.17	3.12	3.06	3.00	2.94	2.88
3.37	3.28	3.18	3.07	3.02	2.96	2.91	2.85	2.79	2.72
3.25	3.15	3.05	2.95	2.89	2.84	2.78	2.72	2.66	2.60
3.15	3.05	2.95	2.84	2.79	2.73	2.67	2.61	2.55	2.49
3.06	2.96	2.86	2.76	2.70	2.64	2.59	2.52	2.46	2.40
2.99	2.89	2.79	2.68	2.63	2.57	2.51	2.45	2.38	2.32
2.92	2.82	2.72	2.62	2.56	2.50	2.44	2.38	2.32	2.25
2.87	2.77	2.67	2.56	2.50	2.44	2.38	2.32	2.26	2.19
2.82	2.72	2.62	2.51	2.45	2.39	2.33	2.27	2.20	2.13
2.77	2.68	2.57	2.46	2.41	2.35	2.29	2.22	2.16	2.09
2.73	2.64	2.53	2.42	2.37	2.31	2.25	2.18	2.11	2.04
2.70	2.60	2.50	2.39	2.33	2.27	2.21	2.14	2.08	2.00
2.67	2.57	2.47	2.36	2.30	2.24	2.18	2.11	2.04	1.97
2.64	2.54	2.44	2.33	2.27	2.21	2.15	2.08	2.01	1.94
2.61	2.51	2.41	2.30	2.24	2.18	2.12	2.05	1.98	1.91
2.59	2.49	2.39	2.28	2.22	2.16	2.09	2.03	1.95	1.88
2.57	2.47	2.36	2.25	2.19	2.13	2.07	2.00	1.93	1.85
2.55	2.45	2.34	2.23	2.17	2.11	2.05	1.98	1.91	1.83
2.53	2.43	2.32	2.21	2.15	2.09	2.03	1.96	1.89	1.81
2.51	2.41	2.31	2.20	2.14	2.07	2.01	1.94	1.87	1.79
2.39	2.29	2.18	2.07	2.01	1.94	1.88	1.80	1.72	1.64
2.27	2.17	2.06	1.94	1.88	1.82	1.74	1.67	1.58	1.48
2.16	2.05	1.94	1.82	1.76	1.69	1.61	1.53	1.43	1.31
2.05	1.94	1.83	1.71	1.64	1.57	1.48	1.39	1.27	1.00

TABLE I continued

Upper tail probability α = .01

Numerator Degrees of Freedom

		1	2	3	4	5	6	7	8	9
	1	4052	4999.5	5403	5625	5764	5859	5928	5982	6022
	2	98.50	99.00	99.17	99.25	99.30	99.33	99.36	99.37	99.39
	3	34.12	30.82	29.46	28.71	28.24	27.91	27.67	27.49	27.35
	4	21.20	18.00	16.69	15.98	15.52	15.21	14.98	14.80	14.66
	5	16.26	13.27	12.06	11.39	10.97	10.67	10.46	10.29	10.16
	6	13.75	10.92	9.78	9.15	8.75	8.47	8.26	8.10	7.98
	7	12.25	9.55	8.45	7.85	7.46	7.19	6.99	6.84	6.72
	8	11.26	8.65	7.59	7.01	6.63	6.37	6.18	6.03	5.91
	9	10.56	8.02	6.99	6.42	6.06	5.80	5.61	5.47	5.35
Denominator Degrees of Freedom	10	10.04	7.56	6.55	5.99	5.64	5.39	5.20	5.06	4.94
	11	9.65	7.21	6.22	5.67	5.32	5.07	4.89	4.74	4.63
	12	9.33	6.93	5.95	5.41	5.06	4.82	4.64	4.50	4.39
	13	9.07	6.70	5.74	5.21	4.86	4.62	4.44	4.30	4.19
	14	8.86	6.51	5.56	5.04	4.69	4.46	4.28	4.14	4.03
	15	8.68	6.36	5.42	4.89	4.56	4.32	4.14	4.00	3.89
	16	8.53	6.23	5.29	4.77	4.44	4.20	4.03	3.89	3.78
	17	8.40	6.11	5.18	4.67	4.34	4.10	3.93	3.79	3.68
	18	8.29	6.01	5.09	4.58	4.25	4.01	3.84	3.71	3.60
	19	8.18	5.93	5.01	4.50	4.17	3.94	3.77	3.63	3.52
	20	8.10	5.85	4.94	4.43	4.10	3.87	3.70	3.56	3.46
	21	8.02	5.78	4.87	4.37	4.04	3.81	3.64	3.51	3.40
	22	7.95	5.72	4.82	4.31	3.99	3.76	3.59	3.45	3.35
	23	7.88	5.66	4.76	4.26	3.94	3.71	3.54	3.41	3.30
	24	7.82	5.61	4.72	4.22	3.90	3.67	3.50	3.36	3.26
	25	7.77	5.57	4.68	4.18	3.85	3.63	3.46	3.32	3.22
	26	7.72	5.53	4.64	4.14	3.82	3.59	3.42	3.29	3.18
	27	7.68	5.49	4.60	4.11	3.78	3.56	3.39	3.26	3.15
	28	7.64	5.45	4.57	4.07	3.75	3.53	3.36	3.23	3.12
	29	7.60	5.42	4.54	4.04	3.73	3.50	3.33	3.20	3.09
	30	7.56	5.39	4.51	4.02	3.70	3.47	3.30	3.17	3.07
	40	7.31	5.18	4.31	3.83	3.51	3.29	3.12	2.99	2.89
	60	7.08	4.98	4.13	3.65	3.34	3.12	2.95	2.82	2.72
	120	6.85	4.79	3.95	3.48	3.17	2.96	2.79	2.66	2.56
	∞	6.63	4.61	3.78	3.32	3.02	2.80	2.64	2.51	2.41

Numerator Degrees of Freedom

10	12	15	20	24	30	40	60	120	∞
6056	6106	6157	6209	6235	6261	6287	6313	6339	6366
99.40	99.42	99.43	99.45	99.46	99.47	99.47	99.48	99.49	99.50
27.23	27.05	26.87	26.69	26.60	26.50	26.41	26.32	26.22	26.13
14.55	14.37	14.20	14.02	13.93	13.84	13.75	13,65	13.56	13.46
10.05	9.89	9.72	9.55	9.47	9.38	9.29	9.20	9.11	9.02
7.87	7.72	7.56	7.40	7.31	7.23	7.14	7.06	6.97	6.88
6.62	6.47	6.31	6.16	6.07	5.99	5.91	5.82	5.74	5.65
5.81	5.67	5.52	5.36	5.28	5.20	5.12	5.03	4.95	4.86
5.26	5.11	4.96	4.81	4.73	4.65	4.57	4.48	4.40	4.31
4.85	4.71	4.56	4.41	4.33	4.25	4.17	4.08	4.00	3.91
4.54	4.40	4.25	4.10	4.02	3.94	3.86	3.78	3.69	3.60
4.30	4.16	4.01	3.86	3.78	3.70	3.62	3.54	3.45	3.36
4.10	3.96	3.82	3.66	3.59	3.51	3.43	3.34	3.25	3.17
3.94	3.80	3.66	3.51	3.43	3.35	3.27	3.18	3.09	3.00
3.80	3.67	3.52	3.37	3.29	3.21	3.13	3.05	2.96	2.87
3.69	3.55	3.41	3.26	3.18	3.10	3.02	2.93	2.84	2.75
3.59	3.46	3.31	3.16	3.08	3.00	2.92	2.83	2.75	2.65
3.51	3.37	3.23	3.08	3.00	2.92	2.84	2.75	2.66	2.57
3.43	3.30	3.15	3.00	2.92	2.84	2.76	2.67	2.58	2.49
3.37	3.23	3.09	2.94	2.86	2.78	2.69	2.61	2.52	2.42
3.31	3.17	3.03	2.88	2.80	2.72	2.64	2.55	2.46	2.36
3.26	3.12	2.98	2.83	2.75	2.67	2.58	2.50	2.40	2.31
3.21	3.07	2.93	2.78	2.70	2.62	2.54	2.45	2.35	2.26
3.17	3.03	2.89	2.74	2.66	2.58	2.49	2.40	2.31	2.21
3.13	2.99	2.85	2.70	2.62	2.54	2.45	2.36	2.27	2.17
3.09	2.96	2.81	2.66	2.58	2.50	2.42	2.33	2.23	2.13
3.06	2.93	2.78	2.63	2.55	2.47	2.38	2.29	2.20	2.10
3.03	2.90	2.75	2.60	2.52	2.44	2.35	2.26	2.17	2.06
3.00	2.87	2.73	2.57	2.49	2.41	2.33	2.23	2.14	2.03
2.98	2.84	2.70	2.55	2.47	2.39	2.30	2.21	2.11	2.01
2.80	2.66	2.52	2.37	2.29	2.20	2.11	2.02	1.92	1.80
2.63	2.50	2.35	2.20	2.12	2.03	1.94	1.84	1.73	1.60
2.47	2.34	2.19	2.03	1.95	1.86	1.76	1.66	1.53	1.38
2.32	2.18	2.04	1.88	1.79	1.70	1.59	1.47	1.32	1.00

TABLE J

Random Digits

```
47505 02008   20300 87188   42505 40294   04404 59286   95914 07191
13350 08414   64049 94377   91059 74531   56228 12307   87871 97064
33006 92690   69248 97443   38841 05051   33756 24736   43508 53566
55216 63886   06804 11861   30968 74515   40112 40432   18682 02845
21991 26228   14801 19192   45110 39937   81966 23258   99348 61219

71025 28212   10474 27522   16356 78456   46814 28975   01014 91458
65522 15242   84554 74560   26206 49520   65702 54193   25583 54745
27975 54923   90650 06170   99006 75651   77622 20491   53329 12452
07300 09704   36099 61577   34632 55176   87366 19968   33986 46445
54357 13689   19569 03814   47873 34086   28474 05131   46619 41499

00977 04481   42044 08649   83107 02423   46919 59586   58337 32280
13920 78761   12311 92808   71581 85251   11417 85252   61312 10266
08395 37043   37880 34172   80411 05181   58091 41269   22626 64799
46166 67206   01619 43769   91727 06149   17924 42628   57647 76936
87767 77607   03742 01613   83528 66251   75822 83058   97584 45401

29880 95288   21644 46587   11576 30568   56687 83239   76388 17857
36248 36666   14894 59273   04518 11307   67655 08566   51759 41795
12386 29656   30474 25964   10006 86382   46680 93060   52337 56034
52068 73801   52188 19491   76221 45685   95189 78577   36250 36082
41727 52171   56719 06054   34898 93990   89263 79180   39917 16122

49319 74580   57470 14600   22224 49028   93024 21414   90150 15686
88786 76963   12127 25014   91593 98208   27991 12539   14357 69512
84866 95202   43983 72655   89684 79005   85932 41627   87381 38832
11849 26482   20461 99450   21636 13337   55407 01897   75422 05205
54966 17594   57393 73267   87106 26849   68667 45791   87226 74412

10959 33349   80719 96751   25752 17133   32786 34368   77600 41809
22784 07783   35903 00091   73954 48706   83423 96286   90373 23372
86037 61791   33815 63968   70437 33124   50025 44367   98637 40870
80037 65089   85919 73491   36170 82988   52311 59180   37846 98028
72751 84359   15769 13615   70866 37007   74565 92781   37770 76451

18532 03874   66220 79050   66814 76341   42452 65365   07167 90134
22936 22058   49171 11027   07066 14606   11759 19942   21909 15031
66397 76510   81150 00704   94990 68204   07242 82922   65745 51503
89730 23272   65420 35091   16227 87024   56662 59110   11158 67508
81821 75323   96068 91724   94679 88062   13729 94152   59343 07352

94377 82554   53586 11432   08788 74053   98312 61732   91248 23673
68485 49991   53165 19865   30288 00467   98105 91483   89389 61991
07330 07184   86788 64577   47692 45031   36325 47029   27914 24905
10993 14930   35072 36429   26176 66205   07758 07982   33721 81319
20801 15178   64453 83357   21589 23153   60375 63305   37995 66275

79241 35347   66851 79247   57462 23893   16542 55775   06813 63512
43593 39555   97345 58494   52892 55080   19056 96192   61508 23165
29522 62713   33701 17186   15721 95018   76571 58615   35836 66260
88836 47290   67274 78362   84457 39181   17295 39626   82373 10883
65905 66253   91482 30689   81313 01343   37188 37756   04182 19376

44798 69371   07865 91756   42318 63601   53872 93610   44142 89830
35510 99139   32031 27925   03560 33806   85092 70436   94777 57963
50125 93223   64209 49714   73379 89975   38567 44316   60262 10777
25173 90038   63871 40418   23818 63250   05118 52700   92327 55449
68459 90094   44995 93718   83654 79311   18107 12557   09179 28408
```

```
24810 61498   24935 86366   58262 44565   91426 86742   61747 79346
75555 42967   02810 16754   08813 40079   62385 21488   38665 94197
49718 05437   73341 17414   91868 24102   76123 67138   43728 43627
04194 83554   98004 14744   63132 75018   75167 24090   02458 78215
66155 24222   91229 63841   03271 56726   36817 51182   94336 20894

10801 76783   05312 30807   40006 04465   70163 17305   46414 76468
40422 75576   82884 27651   58495 87538   23570 66469   46900 95568
95572 12125   12054 17028   03599 73764   48694 85960   89763 58305
57276 49498   88937 08659   46840 83231   75611 94911   28467 67928
80566 94963   83787 87636   89511 64735   86699 66988   91224 72484

44517 55108   17435 33109   60343 46193   66019 43713   24097 52921
55424 87650   13896 90005   99458 20153   86688 13650   75201 79447
80506 78301   97762 16434   62430 28438   13602 63236   81431 75641
03646 54402   75413 39128   82975 73849   27269 73444   26120 06824
14537 53791   43951 51326   33274 54833   80802 66976   04878 35832

01644 33630   71247 59273   07811 33546   88628 06469   86257 39298
39387 94217   77995 53285   13354 84980   83590 63494   06036 18502
74962 49489   54662 93588   50466 55026   62458 06195   07995 71054
21165 45577   46383 38855   21561 89332   94248 09703   78397 38770
58519 95396   73607 72106   76597 85596   99075 39195   99605 66179

46982 79519   22294 15676   83484 98279   79200 02640   22501 43073
58463 67619   18006 05028   32441 83599   28915 05362   21612 64681
43055 00020   39254 68439   27399 24259   04641 50935   07112 55117
84073 38387   14337 90766   60436 65757   57590 17880   13776 35810
93542 37270   09361 62404   74056 52964   67372 81398   01482 97589

54467 20234   52813 85296   14542 73241   74848 39001   97598 76641
43608 42832   93917 67031   50220 94089   64858 27691   16719 99870
64808 01692   46424 64722   87162 06582   01452 14980   17397 07403
78703 93006   59651 48404   82284 66405   89818 00989   56112 78144
14886 70359   32158 30401   20829 22534   88848 07669   25100 48602

69280 61856   78974 91485   01583 11620   53740 32705   80391 56749
99680 99636   54107 79588   90845 21652   58875 13171   68531 18550
01662 21554   63836 41530   21864 81711   68921 61749   36051 78024
67852 69123   14280 17647   65125 82427   61594 32015   93473 05627
13911 67691   97854 89950   40963 06697   82660 69097   65284 49808

95822 09552   65950 34875   64250 41385   80133 70818   09286 30769
44068 24928   27345 34235   44124 06435   06281 43723   97380 76080
99222 66415   71069 62293   77467 35751   22548 23799   96272 58777
08442 61287   72421 35777   61079 42462   17761 94518   98114 74035
14967 60637   32097 28122   87708 19378   93372 23225   38453 80331

27864 15358   16499 91903   62987 98198   15036 23293   68241 44450
99678 16125   52978 79815   85990 18659   00113 93253   49186 25165
89143 79403   22324 54261   97830 42630   48494 09999   69961 39421
34135 31532   42025 83214   83730 28249   25629 11494   70726 45051
72117 97579   36071 29261   89937 78208   23747 56756   37453 51344

19725 76199   08620 22682   52907 25194   84597 93419   95762 14991
96997 66390   27609 41570   17749 23185   24475 56451   91471 33969
44158 67618   15572 95162   95842 08301   11906 68081   40436 58735
33839 40750   18898 61650   09970 47651   41205 65020   33537 01022
53070 61630   84434 05732   18094 71669   41033 82402   16415 83958
```

TABLE J continued

66558 78763	55932 15490	46790 47325	60903 15000	90970 06904
93810 69163	27172 10864	39108 79626	90431 44390	54290 70295
72045 47743	33163 88057	14136 55883	71449 68303	54093 95545
64251 86498	77947 21734	23571 86489	90017 24878	91985 03921
82220 31802	84619 51220	34654 60601	15088 26949	23013 72644
04991 91864	49269 66109	92609 37154	53225 73014	01890 04357
73895 65548	31996 73237	62411 22311	87875 79190	28237 73903
03515 01014	83955 11919	71533 71150	45699 95307	77713 66398
78808 89471	65152 62457	32410 14092	13813 08357	65485 83198
50648 45741	81584 54369	01575 92941	05484 41196	61946 89918
52074 39293	45087 07020	04753 69952	45199 83726	11602 57715
08209 01284	83775 89711	92322 31538	15808 94830	69581 94556
24292 33646	26925 04133	04895 07341	81441 53319	60118 98634
68189 39488	61468 23411	36471 65260	30134 55648	39176 61692
63096 33677	78900 30005	17324 83577	16699 62138	73469 89005
67106 05029	82711 17886	38351 42165	71101 37151	13547 38500
36272 89377	49623 71797	57532 90488	32967 60308	57256 66233
33143 08577	38507 85535	62784 29068	42392 41332	71636 49165
94138 78030	28934 91012	45780 66416	14003 05819	71031 00053
65199 99418	58039 96495	95954 48748	93022 46913	26250 35538
49897 14275	22123 84895	77567 17949	57969 31131	95882 08783
85679 52202	37950 09891	45369 48243	84985 08318	09853 86452
21441 75053	31373 89860	47671 33981	55424 33191	53223 54060
54269 06140	67385 01203	51078 48220	46715 09144	61587 63026
09323 22353	58095 97149	63325 18050	36840 26523	72376 89192
50214 26662	95722 78359	49612 75804	89378 96992	76621 91777
52022 22417	70460 91869	45732 54352	87239 41463	40310 45189
47176 39122	11478 75218	04888 49657	84540 49821	80806 83581
35402 23056	72903 95029	05373 35587	23297 11870	18495 79905
65519 29138	08384 41230	29209 87793	06285 90472	47054 57036
11259 05645	07492 16580	90016 22626	23187 25531	24281 05383
56007 97457	05913 74626	33923 25652	75099 73542	29669 82523
24558 05361	28136 58586	74390 95278	70229 15845	83717 91629
52889 89032	03429 81240	54824 16714	79590 91867	36732 16936
39534 82185	56489 40999	31361 98733	68769 77792	52694 14372
04948 01323	21617 23457	29217 65387	33130 30920	26298 84058
69496 38855	02249 50773	93315 41606	73918 32347	06673 95058
16068 36000	08084 66738	15982 82450	09060 49051	31759 54477
11907 18181	20687 05878	33617 16566	67893 50243	08352 64527
18382 61533	42865 90495	74809 70740	24939 43883	86674 40041
75960 66915	02595 66435	55610 42936	52336 15660	28110 10390
20156 11314	51105 46678	39660 54062	81972 55953	99513 41647
10528 02058	80359 99179	65642 93982	07133 38680	45791 79665
48514 83028	06720 33776	10023 14228	56367 04108	54855 77323
92644 48340	75864 50303	09037 02589	37463 81365	18567 65142
41477 25941	98283 49225	08721 84508	97549 46769	98389 19589
41630 05563	83127 71333	68606 49269	89244 53159	02762 18167
78489 08606	66190 02810	49460 63040	00221 27138	36645 02465
44971 55189	26570 98515	37222 54809	17570 64185	56333 72230
68023 51496	96693 76886	54420 59192	11645 54942	31693 28688

79767	68246	17731	26380	53059	95517	78256	75888	07407	26606
51418	77326	50358	33736	42294	05839	41670	06190	79431	17649
99053	75365	68601	11974	06061	70547	34663	91460	16942	36190
76782	71815	25699	18820	46640	66131	68176	06721	96948	48831
00054	75974	39606	34585	11766	24425	59123	43770	01543	48199
76986	25299	56915	10445	17519	06383	69934	38270	50500	40036
09068	40312	32398	66316	86491	25159	68490	99215	13026	99583
14995	57799	04606	39602	96838	82575	60004	76945	87129	57982
52236	23601	02551	73803	43190	45217	25331	24319	47599	87713
04234	88134	76799	00690	00431	35795	68154	13726	92234	38523
09576	62666	62081	25900	85551	67305	74596	19856	95240	27096
08876	02110	88540	67911	22605	78936	53171	47839	31705	40128
90641	63802	62775	99910	54987	25337	87749	40698	13520	75360
81610	68893	76441	86754	68758	63132	70789	03593	60013	50974
24942	40614	82896	62376	22623	71501	62216	04926	92450	02354
69276	15411	39931	05682	84216	78727	45115	16172	48356	14454
26850	52243	72376	69365	46803	17763	81849	28183	15482	41846
58251	44226	99121	51709	55878	38127	62332	17779	60737	14139
80769	38995	78326	43032	42924	48921	92503	48883	39617	16926
77730	27199	41001	37730	17192	24001	68844	39054	03311	29616
14101	79409	55851	59065	09448	12362	37983	30132	44552	64721
78412	27673	22697	90409	80933	14397	65476	65039	54137	05837
80490	04572	83161	12758	77431	46984	36316	69058	80929	40091
48435	79357	07703	58519	22212	49369	39125	24281	10894	37515
55818	67243	32570	46300	30216	37922	59474	90998	32791	50913
38504	19642	20043	51907	12061	45989	34126	15466	60407	41060
62532	88768	10806	78056	56116	83894	04004	89425	98102	25472
41253	47292	21824	28796	66520	02299	35985	21745	67569	43758
21311	49354	26383	15911	58703	66264	52777	88468	78479	93144
79493	81187	22802	81892	23194	30874	29596	95527	05851	69557
86569	20549	41974	78598	04265	52585	06155	92795	47235	13421
71195	66679	83667	01622	90584	80413	10738	89237	40872	88433
34198	37972	11147	54830	11599	39138	63063	12883	50472	78611
49944	77066	51105	10926	02573	88622	25972	26116	20518	74519
20512	37842	93043	16725	38621	89067	28716	87520	86277	00223
95920	01676	50637	32471	85297	72345	36374	27257	66562	54449
69343	67580	74026	53503	36427	91961	40432	67921	77197	76739
63256	54571	48582	61218	68385	62550	00185	33937	14945	00532
07164	39622	15005	81428	35078	98992	58169	46310	02584	62127
32151	06034	35717	86386	09108	69625	80461	22900	75473	71302
36943	36628	02711	19174	95060	91057	91419	16271	92089	33761
42543	86517	62451	83617	33303	00232	91253	11758	42643	60941
82246	71725	68475	14115	88253	67433	08144	97231	14853	02076
13560	22591	42101	98738	59840	45639	67920	00794	89914	62635
13202	90236	87525	19154	01788	21129	30042	88748	86127	65421
06927	25724	01915	54331	39256	35530	93068	28035	38225	55304
06977	02761	45002	56046	62382	41952	37018	55770	84544	38292
71036	59813	93768	99757	94350	56118	52746	69784	08029	44572
52836	53177	39682	70280	84213	49817	79286	16600	47462	17033
27801	54044	16345	35740	82609	20754	88162	88875	02269	35975

TABLE K

Normal Random Deviates

.052	1.504	−1.350	−1.124	−.521	.515	.839	.778	.438	−.550
−.315	−.865	.851	.127	−.379	1.640	−.441	.717	.670	−.301
.938	−.055	.947	1.275	1.557	−1.484	−1.137	.398	1.333	1.988
.497	.502	.385	−.467	2.468	−1.810	−1.438	.283	1.740	.420
2.308	−.399	−1.798	.018	.780	1.030	.806	−.408	−.547	−.280
1.815	.101	−.561	.236	.166	.227	−.309	.056	.610	.732
−.421	.432	.586	1.059	.278	−1.672	1.859	1.433	−.919	−1.770
.008	.555	−1.310	−1.440	−.142	−.295	−.630	−.911	.133	−.308
1.191	−.114	1.039	1.083	.185	−.492	.419	−.433	−1.019	−2.260
1.299	1.918	.318	1.348	.935	1.250	−.175	−.828	−.336	.726
.012	−.739	−1.181	−.645	−.736	1.801	−.209	−.389	.867	−.555
−.586	−.044	−.983	.332	.371	−.072	−1.212	1.047	−1.930	.812
−.122	1.515	.338	−1.040	−.008	.467	−.600	.923	1.126	−.752
.879	.516	−.920	2.121	.674	1.481	.660	−.986	1.644	−2.159
.435	1.149	−.065	1.391	.707	.548	−.490	−1.139	.249	−.933
.645	.878	−.904	.896	−1.284	.237	−.378	−.510	−1.123	−.129
−.514	−1.017	.529	.973	−1.202	.005	−.644	−.167	−.664	.167
.242	−.427	−.727	−1.150	−1.092	−.736	.925	−.050	−.200	−.770
.443	.445	−1.287	−1.463	−.650	.412	−2.714	−.903	−.341	.957
.273	.203	.423	1.423	.508	1.058	−.828	.143	−1.059	.345
.255	1.036	1.471	.476	.592	−.658	.677	.155	1.068	−.759
.858	−.370	.522	−1.890	−.389	.609	1.210	.489	−.006	.834
.097	−1.709	1.790	−.929	.405	.024	−.036	.580	−.642	−1.121
.520	.889	−.540	.266	−.354	.524	−.788	−.497	−.973	1.481
−.311	−1.772	−.496	1.275	−.904	.147	1.497	.657	−.469	−.783
−.604	.857	−.695	.397	.296	−.285	.191	.158	1.672	1.190
−.001	.287	−.868	−.013	−1.576	−.168	.047	−.159	.086	−1.077
1.160	.989	.205	.937	−.099	−1.281	−.276	.845	.752	.663
1.579	−.303	−1.174	−.960	−.470	−.556	−.689	1.535	−.711	−.743
−.615	−.154	.008	1.353	−.381	1.137	.022	.175	.586	2.941
1.578	1.529	−.294	−1.301	.614	.099	−.700	−.003	1.052	1.643
.626	−.447	−1.261	−2.029	.182	−1.176	.083	1.868	.872	.965
−.493	−.020	.920	1.473	1.873	−.289	.410	.394	.881	.054
−.217	.342	1.423	.364	−.119	.509	−2.266	.189	.149	1.041
−.792	.347	−1.367	−.632	−1.238	−.136	−.352	−.157	−1.163	1.305
.568	−.226	.391	−.074	−.312	.400	1.583	.481	−1.048	.759
.051	.549	−2.192	1.257	−1.460	.363	.127	−1.020	−1.192	.449
−.891	.490	.279	.372	−.578	−.836	2.285	−.448	.720	.510
.622	−.126	−.637	1.255	−.354	.032	−1.076	.352	.103	−.496
.623	.819	−.489	.354	−.943	−.694	.248	.092	−.673	−1.428
−1.208	−1.038	.140	−.762	−.854	−.249	2.431	.067	−.317	−.874
−.487	−2.117	.195	2.154	1.041	−1.314	−.785	−.414	−.695	2.310
.522	.314	−1.003	.134	−1.748	−.107	.459	1.550	1.118	−1.004
.838	.613	.227	.308	−.757	.912	2.272	.556	−.041	.008
−1.534	−.407	1.202	1.251	−.891	−1.588	−2.380	.059	.682	−.878
−.099	2.391	1.067	−2.060	−.464	−.103	3.486	1.121	.632	−1.626
.070	1.465	−.080	−.526	−1.090	−1.002	.132	1.504	.050	−.393
.115	−.601	1.751	1.956	−.196	.400	−.522	.571	−.101	−2.160
.252	−.329	−.586	−.118	−.242	−.521	.818	−.167	−.469	.430
.017	.185	.377	1.883	−.443	−.039	−1.244	−.820	−1.171	.104

.615	−.831	−.944	.739	1.620	−1.353	−1.649	−.767	−.585	−1.235
−.293	1.161	−.507	−1.792	.158	.612	−.065	2.418	.606	−.016
.464	−2.225	.773	.216	1.194	1.484	−1.745	.232	−.215	.171
−1.438	.680	2.035	−2.002	.284	−.349	1.021	1.876	−2.014	.481
.120	−.823	.220	.501	−.573	−.973	−.206	.143	−.577	1.067
.560	−.288	−.409	.641	1.037	−.804	−.143	.735	−.361	.246
2.321	1.368	1.337	.153	−.120	−1.315	1.055	1.227	1.423	−.034
−.355	.376	.099	−.971	.275	−.031	1.034	1.815	−.534	.292
−.182	−1.947	.270	−.567	−1.087	1.078	−1.238	1.120	1.186	.173
.458	−1.388	.636	.479	2.431	1.204	−.040	−1.654	1.345	−1.031
.315	1.859	.499	.954	.222	−.150	−.829	.174	.567	.049
1.043	.144	−1.653	−.438	.527	.394	−.089	.528	−1.737	.126
.320	3.475	.764	.856	1.807	−.554	−.404	−.102	.753	−2.212
.959	.873	−.652	.150	−.735	−1.545	−.749	1.643	.764	−.966
−1.769	−.409	−.477	.195	.549	−.679	.609	−1.422	−.213	.113
.528	.291	1.401	1.533	−.950	−.660	.748	.596	−1.657	2.513
.478	−.591	−.283	−.215	−.910	.147	−.442	.055	−2.818	.250
−.140	−.550	−1.616	1.398	−.568	.396	−.355	−.054	−.082	−1.372
.126	−.086	1.433	−1.008	.405	.264	−1.659	−.503	.455	−.880
−1.070	−1.331	.403	.909	.858	−.949	−1.298	.640	.747	1.044
1.104	−1.204	.281	−1.890	.242	−.593	−.864	−.789	−.233	−.406
−.212	.562	−1.094	.643	−1.765	−1.360	−.123	.414	.137	.631
2.652	−.408	.272	.279	.105	−.395	−.546	−.075	.340	.511
−1.052	−1.864	1.012	.412	−.607	.018	.779	−2.271	−1.862	−1.531
1.061	−.429	−.219	.647	.012	−1.371	.509	.590	1.125	−.373
−1.125	.701	−1.279	−.863	−1.631	−1.323	−.020	−1.402	.114	−2.765
1.038	.117	.225	.120	1.197	1.663	−.841	−1.257	.342	−.292
.139	.031	−2.348	−2.034	−1.004	−.574	.301	.445	−.442	−.664
−.967	.010	1.219	−.974	−1.224	.027	.050	−1.135	−1.725	−.006
1.680	−.866	.152	1.266	−1.241	.954	−1.288	−1.933	1.187	1.004
1.179	−1.542	1.122	−.697	−.524	1.660	.677	.714	.430	−.501
−.839	−.222	2.192	.376	.515	−.751	−.305	−2.091	−1.072	−.361
−.842	−1.738	−1.113	−1.326	−1.970	−.683	1.036	.230	.032	1.813
−1.445	.470	−.007	.081	1.346	−.352	−.813	−.028	−.564	.567
.003	−.258	−.238	−1.008	−.383	−.178	−1.079	.339	−.401	.410
−.206	−1.224	−1.002	1.159	−.456	−.547	.590	−1.224	−1.384	1.580
−.322	−2.374	−1.673	1.348	−.375	1.101	1.801	−.515	−1.247	−.836
2.140	1.230	.048	.411	.417	−.007	−1.573	−1.374	.628	−.586
.287	−.944	1.601	.153	−1.330	.759	−.624	.586	−1.203	.142
1.068	−.662	.877	.188	−.100	−1.809	.024	−1.187	.597	−.291
1.267	−.960	−.530	−.634	1.844	1.942	1.202	−1.084	−1.040	−.764
.500	.123	−.641	.514	−.523	−.389	−1.768	−.445	.685	−1.165
1.396	.469	−1.230	−.473	−1.497	−.263	.999	−.973	−.655	.494
−.179	−.077	−.419	−.578	−2.237	−1.946	−.206	−.999	.918	−.470
−1.581	.880	1.444	−.756	2.018	−1.206	−.202	.675	1.429	−.635
−.719	−.953	−.695	.406	.745	.466	.044	−1.074	1.994	−.279
−.604	2.036	2.659	−1.325	−.476	−.907	.191	.014	1.342	−.063
−.436	−.417	−.068	−.327	−.489	.779	.041	−2.458	−.347	.474
−.128	.501	.406	−.124	−.190	−1.056	.567	−.956	−.292	−.080
.954	−.521	−.030	−.850	2.072	1.593	.466	.193	−.583	−.362

Bibliography

General:

Bates, G. E., *Probability*, Reading, Mass.: Addison-Wesley Publishing Company, Inc., 1965.

Huff, D., *How to Lie with Statistics,* New York: W. W. Norton & Company, Inc., 1954.

Kruskal, W. K., "Statistics, Part I," *International Encyclopedia of the Social Sciences*, Vol. 15, pp. 206–223, New York: Macmillan & The Free Press, 1968.

Moroney, M. J., *Facts from Figures,* Baltimore: Penguin Books, 1953.

Peters, W. S., *Readings in Applied Statistics*, Englewood Cliffs, N.J.: Prentice-Hall, Inc., 1969.

Raiffa, H., *Decision Analysis*, Reading, Mass.: Addison-Wesley Publishing Company, Inc., 1968.

Ruggles, R., and Brodie, H., "An Empirical Approach to Economic Intelligence in World War II," *Journal of the American Statistical Association,* Vol. 42 (1947), pp. 72–91.

Tippett, L. H. C., *Statistics*, 3rd ed., London: Oxford University Press, 1968.

Wallis, W. A., and Roberts, H. V., *The Nature of Statistics*, New York: The Free Press, 1965.

Elementary Treatments of Probability and Statistics:

Blackwell, D., *Basic Statistics,* New York: McGraw-Hill, Inc., 1969.

Hodges, J. L., Jr., and Lehman, E. L., *Basic Concepts of Probability and Statistics,* San Francisco: Holden-Day, 1964.

Mosteller, F., Rourke, R. E. K., and Thomas, G. B., *Probability and Statistics,* Reading, Mass.: Addison-Wesley Publishing Company, Inc., 1961.

Wallis, W. A., and Roberts, H. V., *Statistics: A New Approach*, New York: The Free Press, 1956.

Statistical Methods:

Dixon, W. J., and Massey, F. J., Jr., *Introduction to Statistical Analysis*, 3rd ed., New York: McGraw-Hill, Inc., 1969.

Natrella, M. G., *Experimental Statistics*, National Bureau of Standards Handbook 91, Superintendent of Documents, Washington, D. C., 1963.

Snedecor, G. W., and Cochran, W. G., *Statistical Methods*, 6th ed., Ames, Iowa: Iowa State University Press, 1967.

Statistical Tables:

Harvard University, *Tables of the Cumulative Binomial Probability Distribution*, Vol. 35, Annals of the Computation Laboratory, Cambridge, Mass.

Owen, D. B., *Handbook of Statistical Tables*, Reading, Mass.: Addison-Wesley Publishing Company, Inc., 1962.

Pearson, E. S., and Hartley, H. O., *Biometrika Tables for Statisticians*, Vol. 1, New York: Cambridge University Press, 1954.

RAND Corporation, *A Million Random Digits with 100,000 Normal Deviates,* Glencoe, Illinois: Free Press, 1955.

The Dixon and Massey and the Natrella books contain many statistical tables.

Index

251